Nanotechnology: Lessons from Nature

Discoveries, Research, and Applications

Synthesis Lectures on Engineering, Science, and Technology

Each book in the series is written by a well known expert in the field. Most titles cover subjects such as professional development, education, and study skills, as well as basic introductory undergraduate material and other topics appropriate for a broader and less technical audience. In addition, the series includes several titles written on very specific topics not covered elsewhere in the Synthesis Digital Library.

The Art of Teaching Physics with Ancient Chinese Science and Technology
Matt Marone
2020

Scientific Analysis of Cultural Heritage Objects
Michael Wiescher and Khachatur Manukyan
2020

Case Studies in Forensic Physics
Gregory A. DiLisi and Richard A. Rarick
2020

An Introduction to Numerical Methods for the Physical Sciences
Colm T. Whelan
2020

Nanotechnology Past and Present
Deb Newberry
2020

Introduction to Engineering Research
Wendy C. Crone
2020

Theory of Electromagnetic Beams
John Lekner
2020

The Search for the Absolute: How Magic Became Science
Jeffrey H. Williams
2020

The Big Picture: The Universe in Five S.T.E.P.S.
John Beaver
2020

Relativistic Classical Mechanics and Electrodynamics
Martin Land and Lawrence P. Horwitz
2019

Generating Functions in Engineering and the Applied Sciences
Rajan Chattamvelli and Ramalingam Shanmugam
2019

Nanotechnology: Lessons from Nature–Discoveries, Research, and Applications

Deb Newberry

ISBN: 978-3-031-03750-4   paperback
ISBN: 978-3-031-03760-3   PDF
ISBN: 978-3-031-03770-2   hardcover

DOI 10.1007/978-3-031-03760-3

A Publication in the Springer series
SYNTHESIS LECTURES ON ENGINEERING, SCIENCE, AND TECHNOLOGY

Lecture #18
Series ISSN
Print 2690-0300   Electronic 2690-0327

Nanotechnology: Lessons from Nature

Discoveries, Research, and Applications

Deb Newberry
Newberry Technology Associates

SYNTHESIS LECTURES ON ENGINEERING, SCIENCE, AND TECHNOLOGY #18

ABSTRACT

As long as humans have existed on the planet, they have looked at the world around them and wondered about much of what they saw.

This book covers 21 different phenomena that have been observed in nature and puzzled about for decades. Only recently, with the development of the microscopes and other tools that allow us to study, evaluate, and test these observed phenomena at the molecular and atomic scale, have researchers been able to understand the science behind these observations.

From the strength of a marine sponge found at the depths of the oceans, to the insect-hydroplaning surface of the edge of a plant, to the intricacies of the eyes of a moth, nanotechnology has allowed science to define and understand these amazing capabilities. In many cases, this new understanding has been applied to products and applications that benefit humans and the environment.

For each of the five ecosystems—the ocean, insects, flora, fauna, and humans—the observations, study and understanding, and applications will be covered. The relationship between the more easily observed macro level and understanding what is found at the nanoscale will also be discussed.

KEYWORDS

nanotechnology, lotus leaf, blue morpho, cephalopods, peacock feathers, iridescent, applications, light interactions, undergraduate

To my children: Jennen, Nathan, and Adriane.

They have loved me without measure through all of my trials and challenges and dealt well with their science crazy Mom, even when I embarrass them by trying to explain the "why" of the world to everyone I meet.

May they always look upon the world with a sense of wonder.

Contents

CHAPTER 1

Introduction

As we strive to understand the world around us and continue to improve the ability to observe that world at a smaller and smaller scale, often what is discovered at the nanoscale is based on what is familiar at the macroscale. Therefore, this chapter first reviews how light interacts with materials. This interaction plays a critical role in the explanation of what is observed in nature which can, in turn, lead to new applications at the macroscale.

Next is a very brief reminder of the interaction of water with surfaces.

Section 1.3 presents an introductory overview of antibacterial surfaces found in nature. Although the interaction with bacteria is not the main focus of this book, it is important to be aware that the same nanoscale structures that allow desert beetles to collect water and butterflies to shine with beautiful colors are the same structures that can prevent bacterial adhesion and the resulting harmful effects.

1.1 THE INTERACTIONS OF LIGHT WITH MATTER

The standard images of light interacting with material depict the light reflecting off a surface, such as a mirror, and the light being "bent," refracted, as it enters a different material such as glass or water.

For reflection, light impinging on a surface at a specific angle from the normal (a line perpendicular to the surface) will reflect off the surface at an angle equal to the incident angle as shown in Figure 1.1.

Figure 1.2 is the traditional image depicting light being refracted. This refraction of light follows Snell's law.

Snell's law: law of refraction

$$n_1 \sin \theta_i = n_2 \sin \theta_r,$$

where

n_1 and n_2 are the indices of refraction of the two different mediums.

θ_i and θ_r are the angles of incidence and refraction, respectively, from the normal.

As the materials become more complex, or additional information is desired, other interactions have been defined.

An important interaction of light with material is Bragg's law of reflection which is depicted in Figure 1.3.

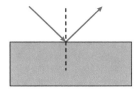

Figure 1.1: Reflection of light off a surface.

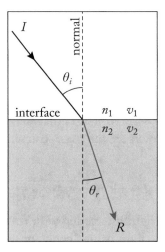

Figure 1.2: The refraction of light.

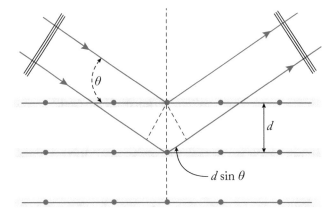

Figure 1.3: Bragg reflection.

Bragg reflection is the foundational principle underlying x-ray diffraction. Where x-rays of a given energy (wavelength) impinge upon a crystal lattice and the x-rays are reflected off of the atoms in the lattice. The atoms in Figure 1.3 are represented by the small dots. The layers of atoms are separated by a distance, d. When the difference in path length, $d \sin \theta$, is equal to the wavelength of the incoming x-ray, the reflected x-rays from the two atoms in different layers are constructive and therefore will show up on a distant screen as a bright spot. When the distance is not an integer value of the spacing, d, then destructive interference will occur, and no light spot will be shown on the screen.

The same effect is exhibited when, instead of specific points, such as the atoms, the incoming x-ray or light wave is reflected from different surfaces within a material. These layers of materials are represented by the shaded horizontal lines in the figure.

This interaction is defined by Bragg's law.

Braggs law: coherent scattering from a crystal structure or two layers.

$$n\lambda = 2d \sin \theta,$$

where

n is an integer.

λ is the wavelength of incident light.

θ is the angle of incidence relative to the plane of the crystal.

Bragg's law, when applied to the interaction of light with multiple layers of material, is often referred to as thin film interference. A four-layer example of *thin film interference* is shown in Figure 1.4. The normal lines are shown as dashed lines in the figure.

The index of refraction, n, is defined as the velocity of the light in a vacuum divided by the velocity of light in the medium. The value of n is equal to 1.0 in a vacuum and is estimated as 1.0 in air. All other mediums, at least at the current time, have an index of refraction greater than 1.0.

When light moves into a medium with a greater index of refraction the light is bent toward the normal. When the light moves into a medium of lower value of n, the light is bent away from the normal. As the light moves through the layers there will be reflected and transmitted portions of the light at each interface (Figure 1.4).

Thin film interference is the phenomena which causes the rainbow of light observed when oil sits on top of water. Light is reflected from the top surface of the oil, the oil and water interface, and the layer of the water on the underlying surface. It is also refracted though the oil and water layers.

You may also notice that as the viewing angle is changed that the color observed changes. This is known as iridescence. When the color observed is dependent upon the angle of observation, the surface or material is termed iridescent.

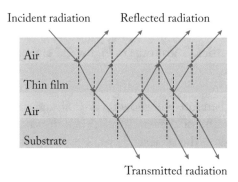

Incident radiation   Reflected radiation

Air

Thin film

Air

Substrate

Transmitted radiation

Figure 1.4: Light interaction in layers of various mediums.

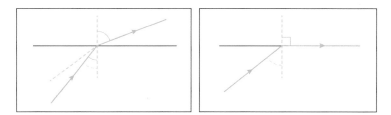

Figure 1.5: The critical angle.

There are circumstances when the index of refraction, the thickness of the material, and the angles involved coincide such that light will not escape from the surface. When this occurs the incident light has reached the "critical angle."

This occurs when light is moving from a more dense or higher index of refraction medium to an environment or material with a lower index of refraction. In this case, the light emerging from the higher n value material to a lower value of n material will be bent away from the normal. There will occur an angle of incidence such that the additional amount of angular bending away from the normal will cause the exiting light to be parallel to the surface. This angle of incidence is defined as the critical angle, as shown in Figure 1.5.

In Figure 1.5 the upper material, above the blue line, has a lower index of refraction than the material below the blue line. The green arrow entering from the lower left represents light impinging on that interface. In the left drawing at the interface the light is bent away from the normal but still exits the structure. In the drawing on the right, the light in impinging on the interface from a slightly larger angle, which, then the angle change due to the change in medium is added to the incident angle, the light is parallel to the interface and cannot be seen.

As shown in Figure 1.5, the interaction of light with material structures takes multiple forms and is dependent on both the nature of the impinging light and the properties of the material with which it is interacting. The region of the electromagnetic spectrum that is of interest

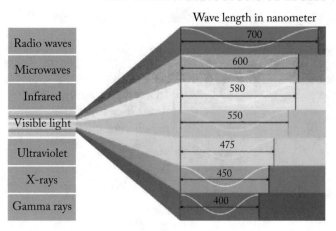

Figure 1.6: The visible portion of the electromagnetic spectrum.

when observing the world around us is the portion of that spectrum in the visible range of light. Light that can be seen by the human eye has wavelengths, λ, in the range of 40–700 nm, as shown in Figure 1.6.

The interaction of light with any material is dependent on the wavelength of the light. Even though in most cases where light interactions are discussed such as Figures 1.1–1.5, the impinging light is shown as a single arrow, usually representing a beam of white light.

For the purposes of understanding what is observed in nature, the wavelength nature of light must be understood and identified. We know that different colors have different wavelengths. We know this inherently: that the interaction of light with a material is dependent upon the wavelength of the light. Otherwise, why do prisms do what they do and why do rainbows appear in the sky?

The interaction of light with a material is dependent upon multiple specific factors and relationships. These factors and relationships are listed in Table 1.1. The dependence of interactions of visible light with a material on the wavelength (color) is a critical aspect of understanding what is observed in nature.

Consider Snell's law of refraction, $n_1 \sin \theta_i = n_2 \sin \theta_r$. This equation does not directly have a wavelength dependence, i.e., there is no λ in the equation. However, n is c/v and $v = f\lambda$! There is the wavelength dependence. Hence, white light entering a prism is separated into different colors because the angle of refraction (Snell's law) is different for different wavelengths.

The same is true for Bragg reflection, or thin film interference. The equations do not specifically include the λ variable, yet the light is only reflected and observed to escape the layers where there is a defined relationship between d, the spacing or thickness of the layers, and the wavelength of the incoming light. Light is only reflected from a layer of material when $d \sin \theta$

Table 1.1: Light factors and relationships

Factor	Designation or Symbol	Description and/or Equation	Application or Comment
Wavelength	λ	In a series of waves, λ is the distance between identical points on adjacent waves	Visible light ranges from 400 nm (blue) to 700 nm (red)
Frequency	f	Rate at which identical points on adjacent waves pass a given point $f = v/\lambda$ v = velocity, m/s	
Energy	E	$E = hf = hv/\lambda$	h = Planks constant
Velocity	v (vee)	v = dist./time = x/t	
Speed of light	c	$c = 1.68 \times 10$ m/s	Vacuum
Angle	θ	Often with a subscript designator θ_i (angle incident) and θ_{ref} (angle reflected or refracted)	Measured with respect to the normal to the surface
Index of refraction	n	$n = c/\text{vel.}_{medium}$	n =1 in air, all other transparent materials $n > 1$

is an integer value of the wavelength. If it is not, then a version of destructive interference will occur, and that wavelength will not be observed.

Consider a simple example of light reflecting off a textured or rough surface, as opposed to the normal flat surface shown in traditional reflection drawings, i.e., Figure 1.1. Three colors of light are shown interacting with a textured surface in Figure 1.7.

In Figure 1.7, the wavelengths comprising the incoming light are interacting with the surface at different locations. As a result, the reflected wavelengths of light heading off to the upper right of the figure will not necessarily be matched up as they were for the incoming set of waves. This results in destructive and constructive interference for the blue and green wavelengths, respectively. For the red light the relationship between the wavelength and the surface roughness is not either an integer or half multiple of the wavelength so a situation of partial destructive interference occurs.

1.2   THE INTERACTION OF WATER WITH SURFACES

Another observation that has garnered the attention of researchers is how various natural surfaces interact with water. One of the primary quantification variables which defines how a surface in-

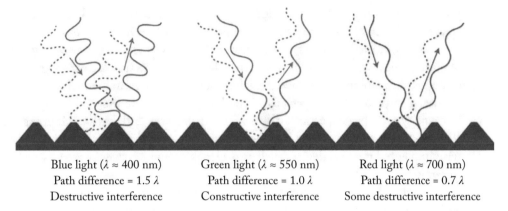

Blue light ($\lambda \approx 400$ nm)	Green light ($\lambda \approx 550$ nm)	Red light ($\lambda \approx 700$ nm)
Path difference = 1.5 λ	Path difference = 1.0 λ	Path difference = 0.7 λ
Destructive interference	Constructive interference	Some destructive interference

Figure 1.7: Three different wavelengths of light interacting with a rough surface.

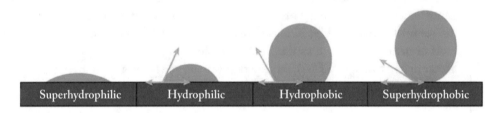

Figure 1.8: Water interactions with a surface.

teracts with water, its wettability, is the angle of attachment between the water and the surface. This angle is the contact angle. Note that although water is the fluid most often considered, the contact angle is an accurate representation of the interaction of any fluid with a surface.

Four categories of the interaction between water and a surface are shown in Figure 1.8. Surfaces on which water forms a spherical droplet are superhydrophobic surfaces, shown at the far right of the figure. The contact angle for a superhydrophobic surface is 140° or greater. The extreme opposite of superhydrophobic is superhydrophilic, shown in the far left image, where the water coats the surface in a thin layer, with a very low contact angle value. The two inner drawings represent the more traditional interactions of hydrophlic and hydrophobic. Often, special structures or chemical coatings are required to create either a superhydophilic or superhydrophobic surface.

1.3   NATURE'S ABUNDANCE OF ANTIBACTERIAL SURFACES

At a macro level, things that can be seen with the human eye, nature abounds with "clean" surfaces, such as the barnacle-free skin of a shark, the pristine surface of a lotus leaf, or the wings of a dragonfly that remain clear of dust and debris. Some of the nanoscale structures that create the observations mentioned above will be discussed in this book.

Many of these surfaces which are clear of artifacts visible to the human eye are also clear of entities that humans cannot see, such as bacteria. Bacteria are microorganisms consisting of a single cell, without an organized nucleus or organelles contained within a cell wall. Bacteria are useful in many ways, especially in the human gut where they aid in digestion, regulate the immune system, protect against disease causing bacteria, and produce vitamins such as B12 and Vitamin K. However, bacteria can also be responsible for many fatal diseases and cause problems with implanted medical devices such as pacemakers and joint replacements such as those of the hip or knee.

Bacterial growth and contamination of implanted devices has been the cause of infections and secondary, follow-up surgeries. Traditionally, chemical cleaning of the surfaces has been used to reduce the occurrences of bacterial infestations. After an infection has occurred, pharmacological therapies are used. Over the last decade, supported by development in the tools of nanoscience that allow living or "wet" biological samples to be studied, effort has been placed on the study of surfaces found in nature that have antibacterial properties. Recall that many of the tools of nanoscience such as the Scanning Electron Microscope (SEM) and the Transmission Electron Microscope (TEM), require that the sample be mounted in a vacuum and often coated with a conductive surface material. This environment is not conducive for the study of biological entities.

Table 1.2 is from a review paper published in the *Journal of Nanobiotechnology*, by Jaggessar et al., which summarizes some of the naturally occurring antibacterial surfaces [3].

As can be seen from Table 1.2, there are many surfaces and structures in nature that exhibit the desired property. Once bacteria enter an environment, it attaches to a surface. Once on the surface, a small number of bacteria cells will secrete a slime layer that enhances the attachment. The bacteria will multiply, and the biofilm, composed of bacteria and attachment slime, will expand to the point of rupture and bacteria will be released into the environment. This scenario is shown in Figure 1.9.

It has been found that surface roughness and topography have a direct correlation to the attachment, or lack of attachment of bacteria. These factors also influence the formation of the biofilm. Other factors such as the chemistry and resulting strength of molecular forces, hydrophobicity, and electrostatic interactions will also impact bacterial adhesion to a surface.

Surfaces with a pillar-type structure, such as that found in the wings of dragonflies and cicadas, will cause the cell wall of an attached bacteria to stretch and sometimes rupture causing the death of the bacteria. A surface which causes bacterial death is called bactericidal. Other surfaces which may prevent or repel cell attachment are termed anti-fouling. An anti-fouling surface may be the result of the topography of the surface or the surface chemistry.

Table 1.2: **Natural antibacterial surfaces** [3]

Natural Surface	Surface	Species	Nano-Texture	Geometry	Contact Angle (°)	Refs.
Plant	Taro leaf	*C. esculenta*	Polygon shape	Bulge: 15–30 μm diam., Papilla: 10–15 μm diam.	159 ± 2	[31, 56]
	Lotus leaf	*N. nucifera*	Micro-size bulge shape	Bulge: 1–5 μm height	142 ± 8.6	[34, 57]
Animal	Gecko skin	*L. steindachneri*	Hair-like nano-structure	4 μm length, top radius of 10–20 nm and submicron spacing	150	[42, 43]
	Shark skin	Spiny dogfish	3D riblet microstructure	Triangular riblets, 100–300 μm width, 15 μm peak radius, 200–500 nm height, 100–300 μm spacing	—	[37]
		C. brachyurous		5 riblets 200–300 μm height, 20–30 μm diameter, 50–80 μm riblet spacing	—	[38]
Insect	Cicada wing	*M. intermedia*	Nano-pillar (conical shape)	Hgt.: 241 nm, diam.: 156 nm, spacing: 165 nm	135.5	[46]
		A. spectabile		Hgt.: 182 nm, diam.: 207 nm, spacing: 251 nm	113.2	[46]
		C. aguila		Hgt.: 182 nm, diam.: 159 nm, spacing: 187 nm	95.7	[46]
		C. maculata		Hgt.: 309 nm, diam.: 97 nm, spacing: 92 nm	76.8 ± 13.9	[45]
		P. scitula		Hgt.: 282 nm, diam.: 84 nm, spacing: 84 nm	91.9 ± 5.9	[45]
		M. hebes		Hgt.: 164 nm, diam.: 85 nm, spacing: 95 nm	78.4 ± 5	[45]
		L. bifuscata		Hgt.: 200 nm, diam.: 90 nm, spacing: 117 nm	81.3 ± 8.3	[45, 58]
		M. conica		Hgt.: 159 nm, diam.: 95 nm, spacing: 115 nm	93.9 ± 8.3	[45]
		M. durga		Hgt.: 257 nm, diam.: 89 nm, spacing: 89 nm	134.8 ± 5.7	[45]
		A. binduvara		Hgt.: 234 nm, diam.: 84 nm, spacing: 91 nm	135.5 ± 5.2	[45, 58]
		M. mongolica		Hgt.: 417 nm, diam.: 128 nm, spacing: 47 nm	123.3 ± 12.7	[45]
		P. radna		Hgt.: 288 nm, diam.: 137 nm, spacing: 44 nm	136.6 ± 5.2	[45]
		D. vaginata		Hgt.: 363 nm, diam.: 132 nm, spacing: 56 nm	141.6 ± 4.5	[45]
		D. vasingra		Hgt.: 316 nm, diam.: 128 nm, spacing: 47 nm	143.6 ± 4.5	[45]
		M. opalifer		Hgt.: 418 nm, diam.: 148 nm, spacing: 48 nm	143.8 ± 6	[45, 58]
		T. vacua		Hgt.: 446 nm, diam.: 141 nm, spacing: 44 nm	144.2 ± 6.8	[45]
		T. impingensis		Hgt.: 391 nm, diam.: 141 nm, spacing: 46 nm	146 ± 2.6	[45]
		C. atrata		Hgt.: 462 nm, diam.: 85 nm, spacing: 90 nm	137.9	[58]
		P. claripennis		Hgt.: 200 nm, base diam.: 100 nm, cap diam.: 60 nm, spacing: 170 nm	147 ± 4.7	[11, 12]
	Dragonfly wing	*S. vulgatum*	Nano-pillar	Height: 80–90 nm, diameter: 150–200 nm	—	[49]
	Butterfly wing	Blue *M. didius*	Scales with aligned micro-grooves	Diameter: 1–2 μm, spacing: 1–2 μm	160	[59]

Examples of anti-fouling topography are surfaces with nanoscale bumps or ridges that prevent the bacteria from connecting with the necessary contact surface area for adhesion. Examples of bactericidal surfaces would be those with stiff hairs or pillars of material close together.

In the following chapters the predominant emphasis is on the nanoscale structures that result in the observed effects at the macro scale. However, many of these nanoscale structures also meet the exact criteria to be anti-fouling or bactericidal surfaces.

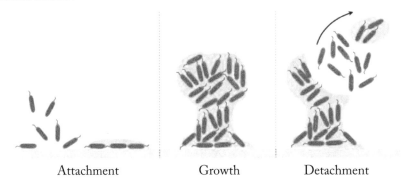

| | Attachment | Growth | Detachment |

Figure 1.9: Bacteria attachment to a surface and the creation of a biofilm.

1.4   REFERENCES

[1] Bhadra, C., Khanh Truong, V., Pham, V. et al. Antibacterial titanium nano-patterned arrays inspired by dragonfly wings. *Scientific Reports*, 5:16817, 2015. https://doi.org/10.1038/srep16817 DOI: 10.1038/srep16817.

[2] Hasan, J. and Chatterjee, K. Recent advances in engineering topography mediated antibacterial surfaces. *Nanoscale*, 7(38):15568–15575, 2015. https://doi.org/10.1039/c5nr04156b DOI: 10.1039/c5nr04156b.

[3] Jaggessar, A., Shahali, H., Mathew, A. et al. Bio-mimicking nano and micro-structured surface fabrication for antibacterial properties in medical implants. *Journal of Nanobiotechnology*, 15(64), 2017. https://doi.org/10.1186/s12951--017-0306-1 DOI: 10.1186/s12951-017-0306-1. 8, 9

[4] MicroOne. Visible light spectrum, color waves length perceived by human eye. Deposit-Photos.com   https://depositphotos.com/stock-photos/light-spectrum.html?filter=all&qview=229295576

CHAPTER 2

From the Sea

INTRODUCTION

The ocean has always provided an environment of beauty, enjoyment, and discovery for scientists and non-scientists alike. Fascinations range from the bioluminescence of waves due to algae and other organisms less than 250 nm in size to gracefully and energy-efficient swimming sharks. Animals that can instantaneously change color and creatures that build strong skeletons at the depths of the sea are among the topics of this chapter.

2.1   CEPHALOPODS

Cuttlefish, octopus, and squid are prominent members of the molluscan class Cephalopoda and are intriguing creatures with their body symmetry (left and right sides), a prominent head, and a set of arms or tentacles. They also have the ability of elaborate camouflage or to extravagantly show off when either is necessary.

This is a result of shutter-like structures called chromatophores. These chromatophores contain a sac of pigment and opening or closing the shutter will expose more or less pigment to the outside world. The chromatophores are coupled with structural reflector cells.

These cells will interact with light either with an iridescent effect or metallic sheen or by multi-layer reflectance. The cells that produce an iridescent response are called iridophores which reflect blue light when viewed directly and red light when viewed more obliquely. These iridophores reflect light by thin film interference due to stacks of thin plates. Iridophores are generally found in squid.

Unlike squid, octopus and cuttlefish often have white areas that are due to leucophores. Leucophores are broadband reflecting structures found as aggregates in the skin of the animal. They contain spherical leucosomes that can be as small as 200 nm in diameter. The leucosomes contain reflectin which is a protein with a high index of refraction. The high index of refraction results in most wavelengths being reflected, hence a white color.

The various structures discussed here are shown in the portions of Figure 2.1. The squid in (a) shows some of the iridescent features and the cuttlefish in (b) adds the white stripes to the mixture. The cuttlefish in (b) and (c) display both camouflage capability and white strips created by the leucophores.

The drawing in (e) is a cross section showing the location of the chromatophores (ch.), iridophores (ir.), and the leucophores (leuc.). Close-up views of the cuttlefish and squid skin are

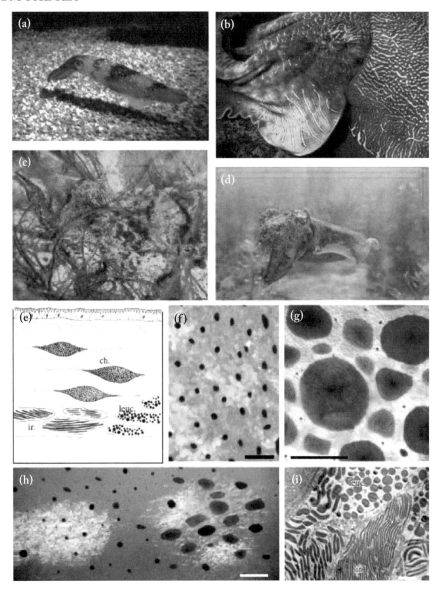

Figure 2.1: Cephalopods with different reflection and color schemes.

shown in (f) and (g), and (h) is a representation of the variety of color and size combinations. The electron microscopy image in (i) presents the cuttlefish iridophore plates and the leucophores spheres. The scale bar in (i) is 1 μm.

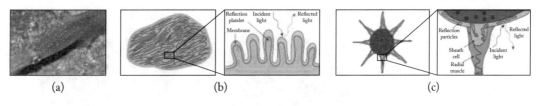

Figure 2.2: Location of reflectin within iridophore and chromatophore.

Reflectin has been found [20] to play a critical role in the coloring of the squid. Figure 2.2 shows the understanding of the location and structure of reflectin within the iridophore and the chromatophore for the squid shown in portion (a) of the figure. Portion (b) shows the reflectin based lamellae under the sheath membrane. Portion (c) shows reflectin particles as part of the chromatophore.

The reflectin protein is a very complex structure showing secondary structures involving β-sheet and α-helical variants. Due to the complexity of the structure, environmental sensitivity, and the multiplicity of nuances, reflectin has been a challenge to study. However, it has been shown that it can be found in the two major organs responsible for the colorations observed in squid and it is found in different forms. It appears that there is a direct correlation between the optical properties of the iridophores and the leucophores and the structural characteristics of the protein reflectin.

Light reflected from an iridophore, and the resulting iridescence, can be modulated by the chromatophore (Figure 2.3). Generally, the chromatophores are located in a layer above the iridophores and can, by their contraction or expansion, change the iridescence (angle dependent color).

The blue-ringed octopus is shown with a close up of the skin and an electron micrograph of the iridophore plates in Figure 2.4.

Leucophores are found in cuttlefish and octopus in addition to iridophores and cause the white patterns. The leucophores reflect ambient light and are found in high density in specific regions of the animal. Figure 2.5 shows various examples of leucophores in cuttlefish. In addition to reflecting white light when it is impinging on the skin surface, the regions of leucophores will reflect any specific wavelength of light. It is proposed that this ability to reflect light of the ambient environment helps with camouflage in the depth of the oceans.

Leucophores occur in partnership with iridophores. Both the plates found in iridophores, and the spheres found in leucophores contain reflectin. The Bragg layers found in iridophores are shown in the right-side drawing of Figure 2.6. The left and center portions of the figure show the physical configuration of the experimental procedure.

Recall that light will only be reflected from multilayered surfaces when there is the appropriate relationship between the wavelength and the thickness of the layers. In addition, the layers must have different values for the index of refraction. When the appropriate relationships

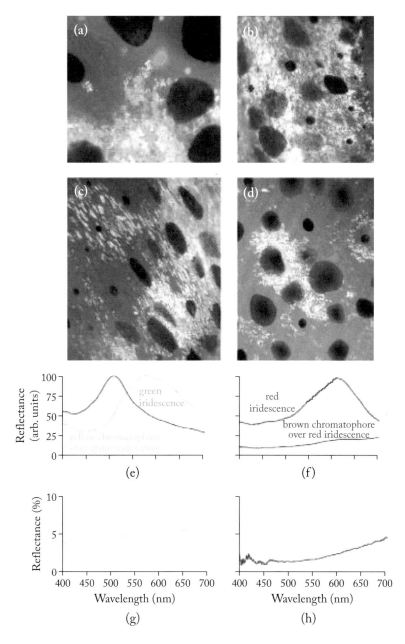

Figure 2.3: Images of chromatophores and iridophores and the reflectance vs. wavelength curves.

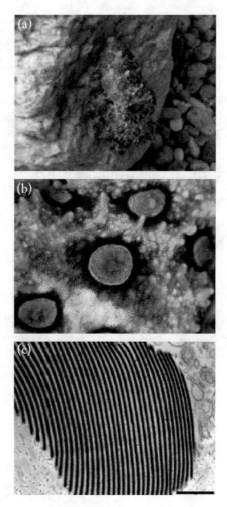

Figure 2.4: Blue-ring octopus, camouflaged, up close, and iridophore plates.

Figure 2.5: Cuttlefish and various images of the leucophores creating the whiteness observed.

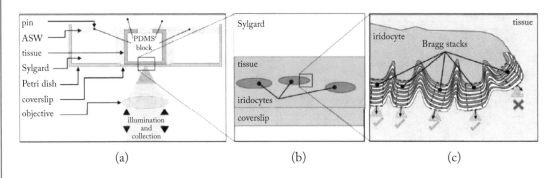

Figure 2.6: Experimental configuration and Bragg layers in squid.

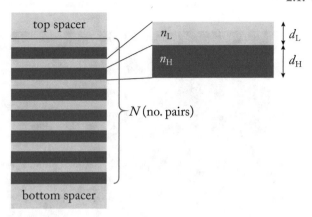

Figure 2.7: Simple representation of a Bragg reflector module.

between these parameters are met, constructive interference will occur and light with a specific wavelength (color) will be reflected. Figure 2.7 shows a simple model of a Bragg reflector used for computer simulations of the iridophore layers. The two layers represent the layers of lamellae filled with reflectin and the spaces in between those layers. The indices of refraction are shown by n_L and n_H, and thicknesses with d_L and d_H. The L and H subscripts represent the low and high values for the index of refraction. Using this physical structure model and experimental measurements of incident light characteristics, (wavelength, angle) and the same characteristics of the reflected light, researchers have been able to create computer generated results that replicate observation.

The detailed structure of an iridophore is presented in Figure 2.8. The right and center columns are SEM images with the center column being a magnification of the first. The scale bars are 20 μm and 5 μm, respectively. The left column shows confocal images of iridophores from the dorsal (top) and ventral side (bottom) of the squid mantle.

In both the plant and animal kingdoms, examples of iridescence can be found, some discussed in this text, however, the tunability and adaptiveness of the iridescence found in the squid is rare. It has been found that the tunability or ability to reflect different colors "on demand" is due to a neurotransmitter acetylcholine (ACh) activation of a signal-transduction cascade. Details of the molecular and cellular drivers behind the process continue to be investigated [17].

A detailed drawing of a side view of the iridophore with the lamellae and a portion of the internal cell structure is given in Figure 2.9. The diagram depicts a potential physical mechanism that could contribute to the modification of the lamellae thickness and index of refraction responsible for the "tailoring" of the reflected wavelength observed.

Similar research efforts have focused on the leucophores, investigating the different types of structures that can result in different degrees of "whiteness" and different shapes [9]. Differences in the densities of the spheres contained within the leucophores contribute to different

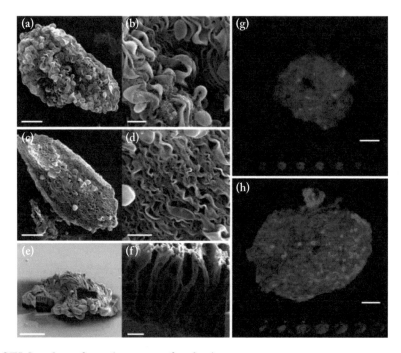

Figure 2.8: SEM and conformal images of iridophores.

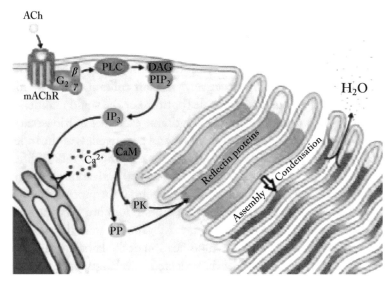

Figure 2.9: Drawing of the proposed mechanism for the modification of the iridophore.

levels of brightness. It has also been found that leucophores and iridophores occur together. Different ratios of the two types of reflector cells will also contribute to different observed responses.

Vocabulary note: Research, investigation, and understanding of the marvel of the color adaptation and changes of the cephalopods continues at a significant pace. One of the discoveries is that the observed color generation is a complex process. The color-generating mechanisms in cephalopods is a combination of multiple cell types, chemical composition, physical attributes, and neural connections and gateways. This complexity and multiple research groups has led to a variety of names for each entity. For example, within different sets of research data, the terms leucophores, leucocytes (not to be confused with leukocytes), and leucosomes may all be referring to nearly the same feature or structure. The same is true for iridophores.

Understanding the mechanisms of the camouflage capability of cuttlefish, squid, and octopus may improve camouflage for military personnel. The color adaptability of the iridophores in squid may provide new approaches to signs and computer screens. Understanding the neural transmittance process which allows the eyes of an octopus to observe the surroundings, transmit that information to the their brain and then send chemical and electrical signals to the necessary cells may support the understanding our of own human brains.

2.2 SHARKS

Cunning, dangerous, fierce—these are words we often associate with the word "shark." Yet understanding several attributes of this creature may prompt a different series of adjectives.

One of the first things that researchers long ago noticed was that the shark remained free of barnacles and other crustaceans that tend to accumulate on whales and other marine animals. Second, the shark was a very efficient swimmer, able to move quickly and smoothly through the water when compared to some other swimmers.

Based on these observations, one of the first aspects of a shark to be investigated was the "skin," as shown in Figure 2.10. What appear to be scales or scale-like structures are modified teeth called placoid denticles. This type of structure is also found on rays and chimaeras. The denticles have an inner tissue component, covered by a layer of dentin and then outer enamel.

Figure 2.10 shows the denticles of the shortfin mako shark in the left image and of the blacktip shark on the right. The scale bars in the images are 100 μm. The ridges in the denticles shown are called riblets or keels. Note that each of the two sharks has differently shaped denticles and number of keels.

A side view of the overlapping denticles of the mako shark is presented in Figure 2.11. The top portion is the crown and is aligned with the direction of water flow over the skin's surface. The scales are attached at the base to the dermis of the skin. The scales between species vary; for example, the shortfin mako shark's scales are 0.18 mm in length whereas the blackfin scales are 0.32 mm in length. As shown in Figure 2.10, the spacing of the keels can also vary. That is, the

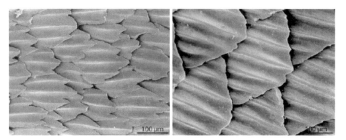

Figure 2.10: Denticles of the shortfin mako shark and the blacktip shark.

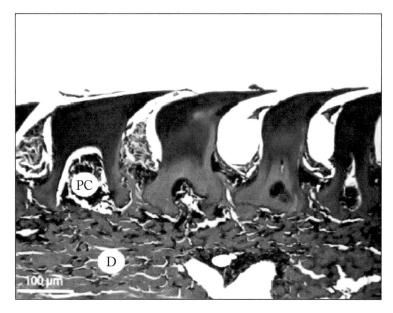

Figure 2.11: The lateral view of the placoid scales showing the crown, base, pulp cavity (PC), and dermis (D).

physical characteristics of the skin of each shark vary, more than likely to be perfected for shark size, environment, and maneuvering requirements.

As noted previously, there is a wide variation among shark species regarding the crown length, number of riblets, and spacing of the riblets. There is also a variation on the scales, dependent upon the location of the scale on the shark as shown in Figure 2.12. The images on the right side are magnifications of the left side images. The green scale bars (left side) are 200 μm and the white bars (right side) are 100 μm.

The riblets or keels play a critical role in reducing drag as the shark swims. The initial inclination may be that a smooth surface should be the best for "gliding" through the water. As

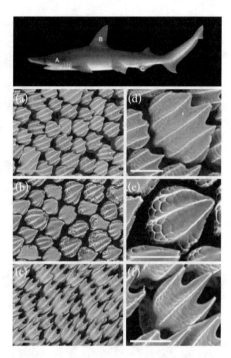

Figure 2.12: The bonnethead shark and ESEM images of the shark scales from different locations on the body.

shown in Figure 2.13, that is not the case. Many experiments have been performed on various test structures which replicate the shark skin. Smoke-filled wind tunnels and colored water pools have been used. The figure shows the air formation for a flat surface (top portion) while the bottom portion represents the riblet structure of the shark scale. The vortices formed in the images are representative of turbulent flow, and what the shark experiences in the ocean. This wind tunnel experience is similar to the work that has been done with golf balls to improve their performance.

The flow direction is into the page and two different velocities are represented. The total shear stress is related to the contact area between the surface and the fluid vortex. The flat surface image shows the created vortices close to the surface and interacting with a large surface area. In the lower portion of the figure the vortices are lifted and only interact with the top surface area of the rib, reducing the shear stress and drag. The overall surface texture, angle of inclination of the riblet and height will also impact the degree of forces acting against the movement of the shark.

Various manufactured replications of the structure of the shark scales have been created for use on boats and other watercraft. A manufactured surface of a shark skin is shown in Figure 2.14. The denticles measure approximately 1.5 mm in length and were manufactured on top of a

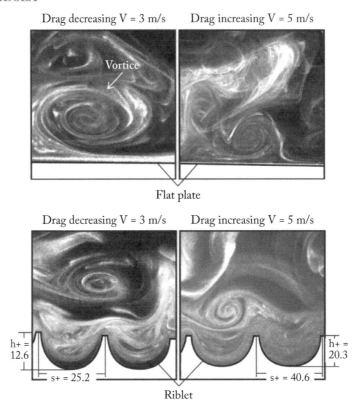

Figure 2.13: Wind tunnel images of the vortices created by a flat surface and a replicated shark scale surface.

flexible surface. Portion (a) of the figure shows the array of denticles both in a concave area where the denticles overlap and the convex segment where they are further apart. Portion (b) shows the sheet in a flat orientation. A single denticle is held on a fingertip in portion (c).

Since the aspect of turbulent flow is not only restricted to water as a fluid, the lessons learned and applied to the water environment are also applicable to air and oil, for example. Figure 2.15 shows an example of the location of a test sheet of shark scale riblet for testing on an airplane wing and also in a test chamber mounted on an airfoil.

Research is showing that the capability of sharks to be such excellent and maneuverable swimmers is due to the size and structure of their scales. Therefore, it stands to reason that because of size and environmental differences that various species of sharks would have different scales and likewise that scales in different parts of the body would vary. Figure 2.16 shows images of the variations of the scales among different species from the same physical location on the animal. The scale bars are 0.5 mm.

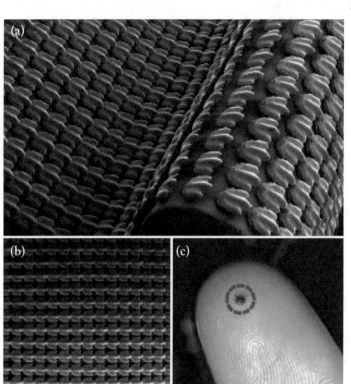

Figure 2.14: A manufactured sheet of shark dentricles.

Although the focus of this section has been on the aspect of drag reduction due to the physical structure of the shark scale, the lack of attached crustaceans has also been studied. It has been found that the texture of the scales, which reduce the drag, also inhibit the attachment of other entities. Since each scale is not a smooth surface, creatures such as mollusks and barnacles cannot attach to enough surface area for adhesion. At the other extreme, although it is possible to attach, small entities such as algae end up separated from each other, hindering replication. This asset of surface structure has also been found to reduce the ability of bacteria to adhere to the surface. Surfaces modeled after the shark skin, when coupled with titanium dioxide nanoparticles, have been found to decrease the ability of microbes to attach and any that do are destroyed [5].

Sharks continue to be fierce denizens of the oceans, yet research is unlocking some of the secrets that lie on the surface (skin) of the shark.

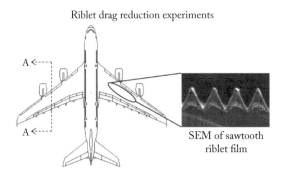

Riblet drag reduction experiments

Airfoil with riblet film covering upper and lower surfaces

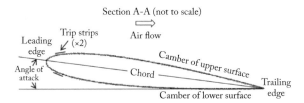

Figure 2.15: Riblet testing in an air fluid flow environment.

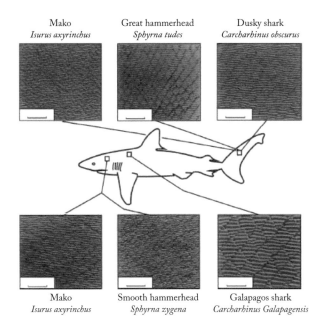

Figure 2.16: Shark scales from similar locations on the body for different species.

Figure 2.17: The Venus flower basket.

2.3   DEEP SEA GLASS SPONGE

Based on our everyday experiences with sponges, our thinking is that they are soft and squishy. That is not quite the case for the deep sea glass sponge found at depths from 300 m and over 4000 m. The glass sponge, class Hexactinellida, is exactly as the name implies, a sea animal with a structure composed of silica, i.e., glass. Figure 2.17 is a photo of one of the most recognizable species, *Euplectella aspergillum*, commonly known as the Venus flower basket.

This beautiful sponge is found in the Pacific Ocean, Indian Ocean, and the Gulf of Mexico. It is a vase-shaped organism, 20–25 cm in length and 2–4 cm in diameter, with a central atrium and a supporting structure of silica and organic fibers called spicules. This unique shape acts as a lifelong home to glass sponge shrimp. A pair of shrimp will breed and offspring will hatch and move on to other sponge homes. Often, the shrimp mature to a size where they cannot escape the web-like structure that covers the top of the sponge "vase" and therefore remain with the sponge in a symbiotic relationship for their lifetime.

In 2005, Aizenberg et al. published work that reviewed the hierarchy of the architecture of the glass sponge from the macro to nano scale [1]. A summary of this architecture at various length scales is presented in Figure 2.18.

With a scale bar of 1 cm shown, in portion (a) of the figure is the entire skeleton of the glass sponge. Portion (b) shows the high-level structural architecture with the checkerboard pattern of squares and diagonal elements. The arrows on the outside of the structure are the orthogonal ridges, the scale bar is 5 mm. Portion (c) shows the deposited silica matrix coating the constituent spicules.

Figure 2.19 is a complement to Figure 2.18 and presents the order of the structure from the nanoscale to the macro scale.

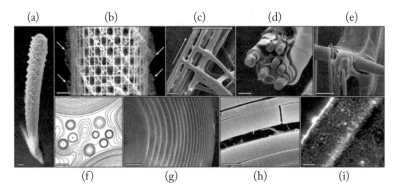

Figure 2.18: Hierarchy of the glass sponge architecture.

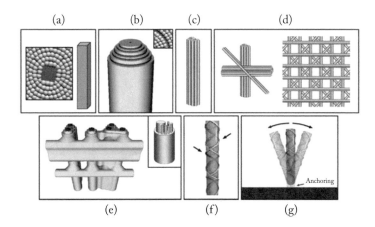

Figure 2.19: Nanoscale to macroscale structure of the glass sponge.

In Figure 2.19 the nanoparticles are represented in (a) (2.18 I—scale bar 500 nm) and shown in (b) arranged in concentric cylinders separated by a thin organic layer, bundled, and (c) (2.18 C scale bar 100 μm, H scale bar 1 μm).

The structure in Figure 2.18 in portion (b) (scale bar 5 mm) is represented here in Figure 2.19 as portion (d), showing the checkerboard pattern created by the vertical and horizontal beams and the diagonal elements which cross every second square. Figure 2.19 (e), with a scale bar of 25 μm, shows the nodes of intersecting spicules with "cement" of a layered silica matrix (inset). Finally, the orthogonal wrapping of ridges on the outside of the resulting structure, (f), and Figure 2.18 (b) helps to maintain the cylindrical shape of the structure. The connection to the soft sea floor is shown in portion (g) of the figure.

The hierarchy of the structural components of the deep sea glass sponge are provided in the chart in Figure 2.20. This chart clearly defines the interdependencies of structures at the

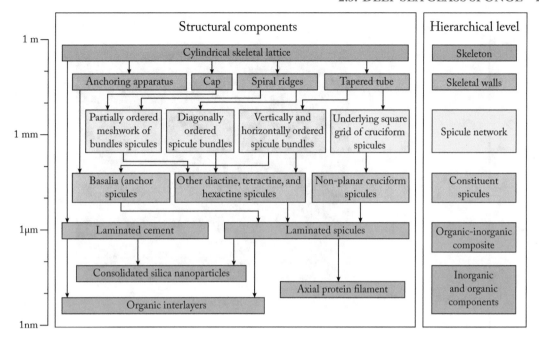

Figure 2.20: Hierarchical organization of structures in the glass sponge.

different length scales, each of which plays a role in the integrity and mechanical properties of this animal.

Continuing work [25] has resulted in characterizing the nanoparticle diameters of 50–200 nm as shown in Figure 2.21 and the spicule structure of a central solid core with concentric circular layers, Figure 2.22.

The deep sea glass sponge is tethered to the ocean floor by anchor spicules. These spicules have a barbed structure with a crown at the farthest end, as shown in portion (c) of Figure 2.23. The dotted lines in (c) show the assumed force direction on the spicule when embedded in the sea floor. Portion (d) of the figure is a cross section of the spicule and designates the outer and inner radii and the solid core.

Just as not all of the external structures of spicules are the same in a specific sponge, as found in the deep sea glass sponge, the internal structure of spicules will vary between different sponges. Figure 2.24 represents the spicule structure of the now familiar glass sponge in portions (a), (b), and (c) of the figure. Portions (d), (e), and (f) are images of the *Tethya aurantia* sponge, known as the golf ball or orange puffball sponge. This sponge is about 10 cm in diameter and is found in the Mediterranean Sea and the North Atlantic Ocean usually in intertidal and subtidal depths to a maximum of 440 m. The golf ball sponge also has spicules as shown in e of the figure with a scale bar of 125 μm. Based on the scale bar, the entire length of the spicule would be

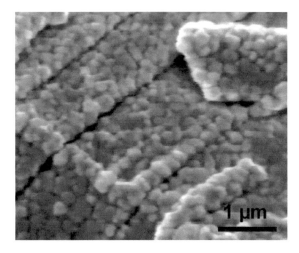

Figure 2.21: SEM image of the silica nanoparticles.

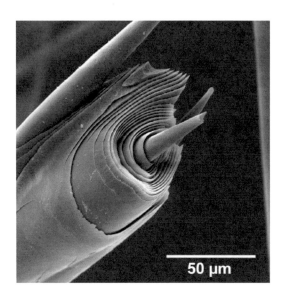

Figure 2.22: SEM image of the cross section of the concentric layered structure of a spicule beam.

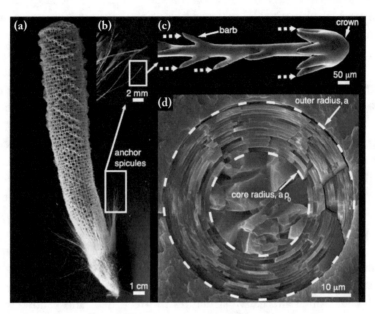

Figure 2.23: The structure of the anchoring spicules of the glass sponge.

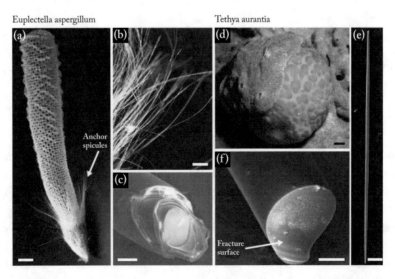

Figure 2.24: Spicule structure comparison between the glass sponge and the golf ball sponge.

1850 μm which is 1.85 mm. Portion (f) of Figure 2.24 is the cross section of the orange puffball sponge which shows a solid material lacking the concentric ring structure. Hence, although the spicule appears to be a common physical aspect among sponges they are not of similar size or structure, alluding to the fact that the functions of each are different.

The skeleton of a glass sponge remains long after the animal has died and remains strong in structure, often being found intact in the ocean. This observation has led many researchers to study, analyze, and test various aspects of the glass sponge structure. One of the architectures studied has been the macro-level structure with the vertical and horizontal beams with diagonal cross bars. Evaluation of this specific structure and quantifying the strength of the structure may lead to better designs of systems which require those properties, such as airplanes, without a weight penalty.

Figure 2.25 shows the biological structure ((a), (b), and (c)) and the translation of that design to modeling parameters, in (d) and (e). Some of the geometric parameters of interest are the width of the "beams" and the separation distances as shown.

Variations on a theme are always appropriate for scientific research with alternate models selected and simulated for comparison to the true design under evaluation. This approach is shown in Figure 2.26 where the upper portion of the figure shows one, the true structure shown in blue and three alternative structures in green, red, and yellow, respectively. Each of these structures was exposed to strain with the results presented in portion (e). The top portion is with no strain applied and the bottom row shows an applied strain. The graph in portion (f) shows the experimental and simulated stress vs. strain response of the four structures. The results show that the architectural design of the glass sponge exhibits better mechanical performance than the other designs.

Animals living in the extreme depths of the oceans have been discovered to include a plethora of structural design approaches to survive and thrive in those environments. Understanding the structures will lead to stronger vehicles and buildings, as well as reducing weight and potentially material costs.

2.4   ABALONE

Like many ocean creatures, the abalone and other members of the mollusk phylum hold the attention and curiosity of humans. Of special interest is the significant strength of the shells of the creatures and the inner beauty that is representative of them. Figure 2.27 is a photo of the red abalone outside and inside. The outside is a typical, non-descript, rough, dull shell. The inside, however, is shiny, smooth and covered in a multi-colored layer of nacre.

As was seen with the deep sea glass sponge, there are substantial aspects of the nanoscale structures that flow through the hierarchy to the macro level observed attributes. Figure 2.28 shows the hierarchy beginning with the atomic-level structures of the calcium carbonate ($CaCO_3$) and the chitin, ($C_8H_{13}O_5N)_n$, a long-chain polymer (functionally comparable to the protein keratin). Chitin is often the organic layer that serves as the "flexible" mortar between the

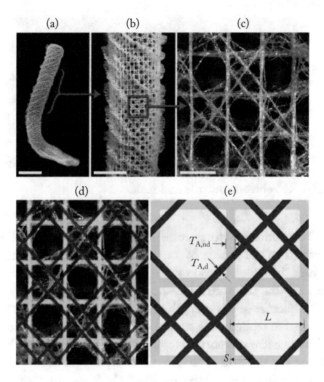

Figure 2.25: Macro-scale skeletal structure of the beams in the deep sea glass sponge and simulation model.

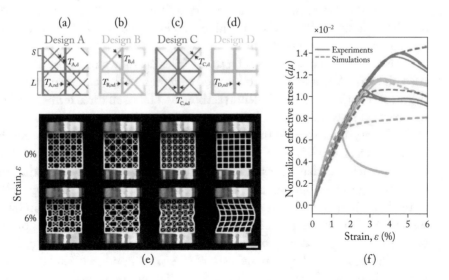

Figure 2.26: Four models for the evaluation (test and simulation) of the structural design of the glass sponge.

Figure 2.27: Red abalone outside and inside.

tablets or bricks in these structures. As a flexible material it will absorb some of the force that may be exerted on the outside of the shell. The calcium carbonate structure can form a mineral bridge, where the mineral breaks through the growth-preventing protein layer as shown in (II) of Figure 2.28. The tiles formed as shown in (III) of the figure are represented more closely in Figure 2.29. Because of the strength of the nacre in the shell, different manufacturing processes have been used to replicate the structure. Portion (b) of the figure is the structure of the tiles in the red abalone nacre and portion (c) is a manufactured nacre. The arrows show the direction of stress applied and the shaded sections focus on regions of slippage along the red lines.

Although the organic layer present in between the tiles or tablets in the nacre region is flexible and serves to absorb some of the force placed on the shell of the animal it may also represent a "slippery" entity that could result in sliding of the layers of tablets. The organic layer, approximately 20-nm thick, is shown very well in Figure 2.30. The holes that are seen in the membrane layer are due to the stretching. Other images show that the organic layer is very soft and not firmly attached to the tiles.

Note: various researchers refer to the "tiles" as shown in Figure 2.28 as tablets or platelets. Also, the term "argonite platelets" is used to represent the same structures. Aragonite is one of the three crystalline forms of calcium carbonate. Hence all three terms; tiles, tablets and platelets may be used by various research groups to describe the same structure.

Observations showed that the nacre tiles did not slide across one another when force was applied. Several models were proposed for the observed results which did not include sliding. One approach was that the asperities or bumps on the surface of the tiles rubbed against each other and prevented sliding. The asperities may be a result of the non-uniform organic layer, or the holes in that layer that allow the mineral components to build up into bumps on the tile surface or remnant of larger mineral bridges between tiles. Another possibility is that the organic layer itself has a viscoelastic property which increases the tensile strength when it is stretched as would occur in a slippage environment. A final option is that the lack of slippage is due to mineral bridges that fracture during the initial force and then slide against the tiles and each other increasing the kinetic friction and reducing slippage. More than likely it is a combination of

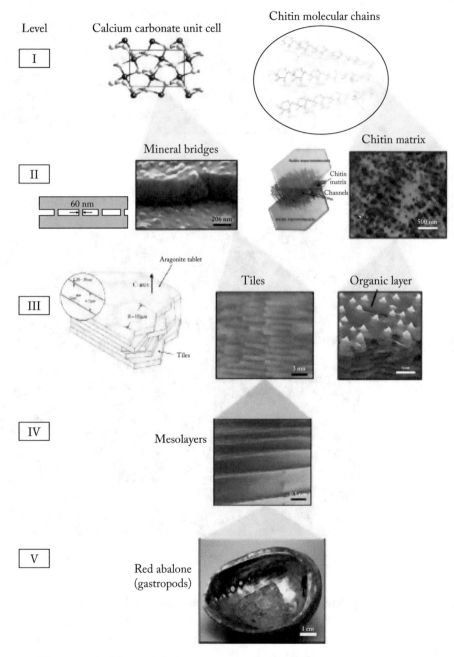

Figure 2.28: Hierarchy of the red abalone.

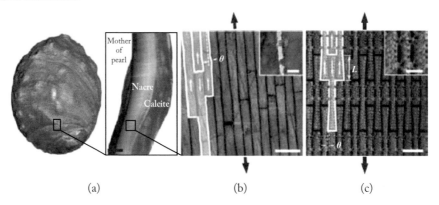

(a) (b) (c)

Figure 2.29: Red abalone tile structure in natural nacre and manufactured nacre.

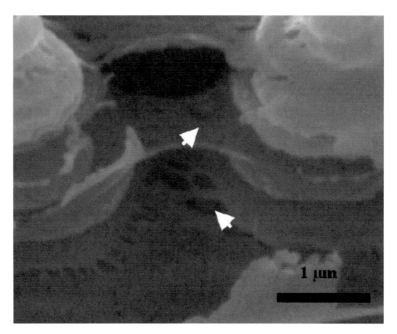

Figure 2.30: SEM image of tiles with the stretched organic layer visible.

all three of the proposed causes. Figure 2.31 includes SEM images of the contributing structures and drawings of the possible explanations for the observed lack of slippage of the tiles.

Figure 2.32 shows tiles that were subjected to a tensile test. Portions (a) and (b) show the organic layer hanging between the tiles. The tile surface and organic layer networks are shown in (c) and (d). The region marked (A) in (d) shows the fabric of the organic layer and region (B) is the mineral.

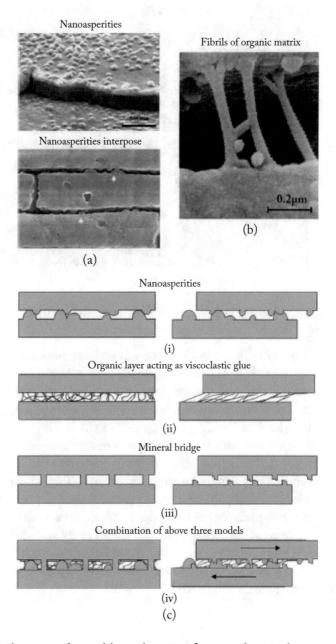

Figure 2.31: Contributors and possible explanation for non-slippage between tiles.

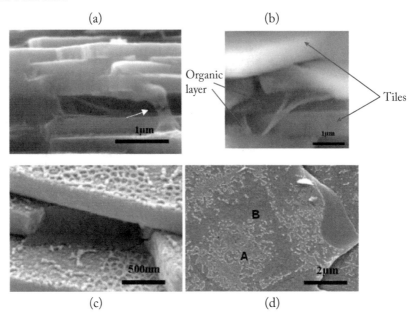

Figure 2.32: SEM images of tiles, organic layer, and asperities.

The content of this section has focused on the architectural nature of the nanoscale structures of the abalone shell, in particular the tile configuration and properties. In addition to this level of research, individual tiles or platelets have been studied [11]. The platelets are closely packed together and composed of argonite with sizes in the range of 5–10 μm. Argonite, being a crystalline form of $CaCO_3$, is composed of nanograins approximately 38 nm in diameter. The nanograins are embedded in a biopolymer matrix approximately 4-nm thick. The result of this work shows that there are large variations (over tens of GPa) in the elastic modulus over distances as small a few hundred nanometers of area. This is believed to be due to the different crystal organizations that can occur due to the nanoparticles that make up the platelets.

The creatures in the ocean never fail to amaze researchers and non-researchers alike. The abalone and related mollusks astound with their beauty and structural integrity. Figure 2.33 is a schematic showing the cross section of the red abalone. Each layer, segment, and constituent has provided insight and understanding that will benefit areas such as medicine, construction, and composite materials. And each layer contains secrets yet to be discovered.

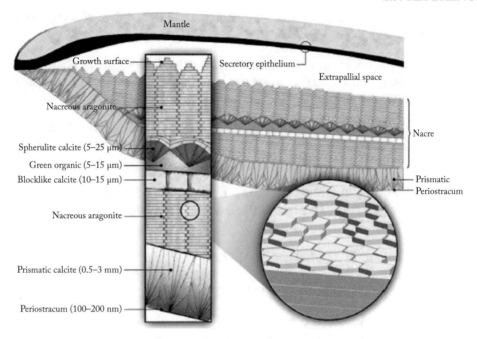

Figure 2.33: **A drawing of the nacre structural pieces in the abalone shell anatomy.**

2.5   REFERENCES

[1] Aizenberg, J., Weaver, J., Thanawala, M., Sundar, V., Morse, D., and Fratzl, P. Skeleton of Euplectella sp: Structural hierarchy from the nanoscale to the macroscale. *Science*, 309:275–278, 2005. DOI: 10.1126/science.1112255. 25

[2] Bixler, G. D. and Bhushan, B. Fluid drag reduction with shark-skin riblet inspired microstructured surfaces. *Advanced Functional Materials*, 23:4507–4528, 2013. https://doi.org/10.1002/adfm.201203683 DOI: 10.1002/adfm.201203683.

[3] Brian, D. and Bharat, B. Shark-skin surfaces for fluid-drag reduction in turbulent flow: A review. *Philosophical Transactions of the Royal Society A*, 368:4775–4806, 2010. http://doi.org/10.1098/rsta.2010.0201 DOI: 10.1098/rsta.2010.0201.

[4] Demartini, D., Krogstad, D., and Morse, D. Membrane invaginations facilitate reversible water flux driving tunable iridescence in a dynamic biophotonic system. *Proc. of the National Academy of Sciences of the United States of America*, 110, 2013. DOI: 10.1073/pnas.1217260110.

[5] Dundar-Arisoy, F., Kolewe, K. W., Homyak, B., Kurtz, I. S., Schiffman, J. D., and Watkins, J. J. Bioinspired photocatalytic shark-skin surfaces with antibacterial

and antifouling activity via nanoimprint lithography. *ACS Applied Material Interfaces*, 10(23):20055–20063, 2018. DOI: 10.1021/acsami.8b05066. 23

[6] Espinosa, H., Juster, A., Latourte, F. et al. Tablet-level origin of toughening in abalone shells and translation to synthetic composite materials. *Nature Communications*, 2:173, 2011. https://doi.org/10.1038/ncomms1172 DOI: 10.1038/ncomms1172.

[7] Fernandes, M. C., Aizenberg, J., Weaver, J. C. et al. Mechanically robust lattices inspired by deep-sea glass sponges. *Nature Materials*, 20:237–241, 2021. https://doi.org/10.1038/s41563--020-0798-1 DOI: 10.1038/s41563-020-0798-1.

[8] Ghoshal, A., Demartini, D., Eck, E., and Morse, D. Optical parameters of the tunable bragg reflectors in squid. *Journal of the Royal Society, Interface*, 2013. DOI: 10.1098/rsif.2013.0386.

[9] Hanlon, R. T., Mäthger, L. M., Bell, G. R. R., Kuzirian, A. M., and Senft, S. L. White reflection from cuttlefish skin leucophores. *Bioinspiration and Biomimetics*, 13(3):035002, March 20, 2020. DOI: 10.1088/1748-3190/aaa3a9. 17

[10] Lang, A., Habegger, M. L., and Motta, P. Shark skin drag reduction. Bhushan, B. (Eds.), *Encyclopedia of Nanotechnology*, Springer, Dordrecht, 2012. https://doi.org/10.1007/978--90-481-9751-4_266 DOI: 10.1007/978-94-007-6178-0.

[11] Li, T. and Zeng, K. Nanoscale elasticity mappings of micro-constituents of abalone shell by band excitation-contact resonance force microscopy. *Nanoscale*, 6, 2013. DOI: 10.1039/c3nr05292c. 36

[12] Li, X., Chang, W.-C., Chao, Y. J., Wang, R., and Chang, M. Nanoscale structural and mechanical characterization of a natural nanocomposite material—the shell of red abalone. *Nano Letters*, 4(4):613–617, 2004. http://dx.doi.org/10.1021/nl049962k DOI: 10.1021/nl049962k.

[13] Mäthger, L. M., Denton, E. J., Marshall, N. J., and Hanlon, R. T. Mechanisms and behavioural functions of structural coloration in cephalopods. *Journal of the Royal Society Interface*, 6(2):S149–63, April 6, 2009. DOI: 10.1098/rsif.2008.0366.focus.

[14] Meyers, M. A., Lin, A. Y., Chen, P. Y., and Muyco, J. Mechanical strength of abalone nacre: Role of the soft organic layer. *Journal of the Mechanical Behavior of Biomedical Materials*, 1(1):76–85, 2008. DOI: 10.1016/j.jmbbm.2007.03.001.

[15] Monn, M., Weaver, J., Zhang, T., Aizenberg, J., and Kesari, H. New functional insights into the internal architecture of the laminated anchor spicules of Euplectella aspergillum. *Proc. of the National Academy of Sciences of the United States of America*, 112, 2015. DOI: 10.1073/pnas.1415502112.

[16] Monn, M. A., Vijaykumar, K., Kochiyama, S. et al. Lamellar architectures in stiff bio-materials may not always be templates for enhancing toughness in composites. *Nature Communications*, 11:373, 2020. https://doi.org/10.1038/s41467--019-14128-8 DOI: 10.1038/s41467-019-14128-8.

[17] Morse, D. E. and Taxon, E. Reflectin needs its intensity amplifier: Realizing the potential of tunable structural biophotonics. *Applied Physics Letters*, 117:220501, 2020. https://doi.org/10.1063/5.0026546 DOI: 10.1063/5.0026546. 17

[18] Sun, J. and Bharat B. Hierarchical structure and mechanical properties of nacre: A re-view. *RSC Advances*, 2:7617–7632, 2012. https://doi.org/10.1039/C2RA20218B DOI: 10.1039/c2ra20218b.

[19] Thompson, J. B., Paloczi, G. T., Kindt, J. H., Michenfelder, M., Smith, B. L., Stucky, G., Morse, D. E., and Hansma, P. K. Direct observation of the transition from calcite to aragonite growth as induced by abalone shell proteins. *Biophysical Journal*, 79(6):3307–3312, 2000. https://doi.org/10.1016/S0006--3495(00)76562-3 DOI: 10.1016/s0006-3495(00)76562-3.

[20] Umerani, M. J., Pratakshya, P., Chatterjee, A., Cerna Sanchez, J. A., Kim, H. S., Ilc, G., Kovačič, M., Magnan, C. N., Marmiroli, B., Sartori, B., Kwansa, A. L., Orins, H., Bartlett, A., Leung, E. M., Feng, Z., Naughton, K. L., Norton-Baker, B., Phan, L., Long, J., Allevato, A., Leal-Cruz, J. E., Lin, Q., Baldi, P., Bernstorff, S., Plavec, J., Yingling, Y. G., and Gorodetsky, A. A. Structure, self-assembly, and properties of a truncated reflectin variant. *Proc. of the National Academy of Sciences of the United States of America*, 117:32891–32901, 2020. DOI: 10.1073/pnas.2009044117. 13

[21] Weaver, J., Aizenberg, J., Fantner, G., Kisailus, D., Woesz, A., Allen, P., Fields, K., Porter, M., Zok, F., Hansma, P., Fratzl, P., and Morse, D. Hierarchical assembly of the siliceous skeletal lattice of the hexactinellid sponge Euplectella aspergillum. *Journal of Structural Biology*, 158:93–106, 2007. DOI: 10.1016/j.jsb.2006.10.027.

[22] Wen, L., Weaver, J. C., and Lauder, G. V. Biomimetic shark skin: Design, fabrication, and hydrodynamic function. *Journal of Experimental Biology*, 217:1656–1666, 2014. https://doi.org/10.1242/jeb.097097 DOI: 10.1242/jeb.097097.

[23] Wikimedia Commons. File:Euplectella aspergillum Okeanos.jpg—Wikimedia Com-mons, 2020. https://commons.wikimedia.org/w/index.php?title=File:Euplectella_aspergillum_Okeanos.jpg&oldid=497144259

[24] Wikimedia Commons. File:AbaloneInside.jpg. 2020. https://commons.wikimedia.org/w/index.php?title=File:AbaloneInside.jpg&oldid=487956978

[25] Woesz, A., Weaver, J., Kazanci, M., Dauphin, Y., Aizenberg, J., Morse, D., and Fratzl, P. Micromechanical properties of biological silica in skeletons of deep-sea sponges. *Journal of Materials Research*, 21, 2006. DOI: 10.1557/jmr.2006.0251. 27

CHAPTER 3

The Insect World

INTRODUCTION

This chapter addresses the science that has evolved to understand the observed phenomena in insects and apply it to multiple applications.

The rainbow of colors observed in the wings of a butterfly are often a result of the structure of the scales on the wing rather than a chemical dye or pigment. The strength of some beetles is due to the design of the exoskeleton rather than the material itself. The anti-reflective eyes of moths have led to improved material surfaces including solar panels.

In this chapter, the first section will cover an investigation of one of the most studied groups of insects: butterflies. Section 3.1 will begin with the members of the genus Morpho, of which there are multiple species. Next, the genus Papilio will be covered. Then, another unique butterfly, the Glasswing, is investigated. In each of these insects, it is the unique nanoscale structure of the wings that results in the observed optical properties.

Two beetles are next investigated, one for its unusual water-gathering methods, the Namib desert beetle, and the other for its amazing strength, the Ironclad Beetle. Finally, the eyes of moths and their particular structure as well as the chemistry and three-pronged nanotechnology-based approach that provides the glow in fireflies conclude the chapter.

3.1   BUTTERFLIES

For centuries humans have been intrigued, amazed, and in awe of the beautiful gracefully flying-butterflies.

Butterflies and moths are members of the Lepidoptera order in the animal kingdom. Lepis comes from the Greek word for "scale" and pteron meaning "wing." This implies that very early on the scale structure of the wings of butterflies and moths was known. Exactly how this was known by those early biologists is not clear and perhaps they envisioned scales by observing the material that remained on their fingers after grabbing a butterfly.

In any event, what we have been able to learn in the last several decades about butterflies, especially the structure of their wings, is because of Atomic Force Microscopes (AFMs) and Scanning Electron Microscopes (SEMs), the tools of nanotechnology that allow observation of our world at the nanoscale.

In the early 2000s, researchers used these tools to begin to unravel the mysteries of the colors on the wings of butterflies and moths.

Figure 3.1: Photograph of the Blue Morpho butterfly and an SEM image of the nanoscale wing structure.

3.1.1   GENUS MORPHO

Of initial interest because of its beauty and size, is what is commonly referred to as the Blue Morpho butterfly. The designation of blue morpho has been applied in a general sense to the multiple species that comprise this genus. These butterflies are some of the largest known, and are found predominantly in Central and South America.

One of the first Blue Morpho butterflies studied was the M. Sulkowskyi butterfly. Researchers were amazed in the early 2000s when it was discovered that a nanoscale physical structure existed in the scales of the wings that could be the reason behind the brilliant shining blue rather than chemical pigmentation or dyes which often create the observed color in insects.

The butterfly (M. Sulkowskyi) and the nanoscale structure discovered are shown in Figure 3.1, in which the brilliant blue is obvious. The SEM image shows the nanoscale structure of the ridge, the center vertical spine, the tapered laminae extending out of the ridge, and the "micro rib" structure. The micro rib structures are smaller vertical structures that run in between the rows of laminae. In this butterfly, the laminae are composed of chitin, a biological collagen

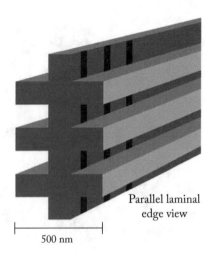

Parallel laminal
edge view

500 nm

Figure 3.2: Drawing of the tree structure from an angle.

material and the spaces are air filled. To get an overall sense of the scale of these structures, consider that each scale on the wing has about 70 ridges. Each ridge is less than 1 μm and they are parallel to each other. The ridges are wider at the base, around 800 nm, and thinner at the top (400 nm). The lamellae are around 80-nm thick and have a vertical spacing of 150 nm. The total height of the ridge is around 2 μm. As is often found in nature there will be variations between species.

Figure 3.2 is a drawing of the nanoscale structure viewed at an angle and Figure 3.3 shows the structure variations and dimensions of the tree-like structures.

The units are nm in Figure 3.3 that shows both the original structure and the replication using a photolithographic process. Note that in the natural structure the lamellae are both in alternating locations on the ridge and in parallel.

For this particular structure it is the combination of the "Christmas tree-" like structure, the alternating laminae layers, and an offset between the neighboring ridges that result in the beautiful iridescent blue color observed. This observed color is independent of the viewing angle and as such this structure has potential for use as a sensor for vapor composition.

Also of importance is the fact that the ridges of the structure form long lines that appear parallel when viewed from the top, as shown in Figure 3.4, third image from the left, yet the ridge structures are somewhat disarrayed in the vertical direction, as shown in the far-right image.

Recall that the interaction of light with a material and the observed result of that interaction is dependent upon multiple factors. These factors include material composition. In materials that are transparent to the light, the critical factor is the index of refraction, n. The value of n which is a ratio of the velocity of light in a vacuum, c, to the velocity of the light in the medium. The value of n will determine how the light is impacted when it enters the medium and whether

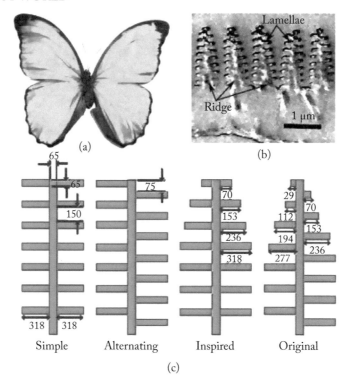

(a)

(b)

(c)

Simple   Alternating   Inspired   Original

Figure 3.3: Blue Morpho structure and variations in organization and dimensions, shown in nm.

Figure 3.4: Images with continuing magnification left to right showing in the first three images a view from the top of the scale and the fourth showing a side view.

it is refracted and/or absorbed within the medium. The interaction of light with a material is also dependent upon the physical structure of the material, such as spacing between and the dimensions of any materials making up the structure as well as the total thickness.

Therefore, the interaction of light in the structure of the Blue Morpho butterfly's wings is dependent upon the material that the ridge, laminae, and ribs are made of, the dimensions of each of those entities, the spacings of the individual ridges, and the "material" that is in the spaces of the structure. In nature this filling material is air.

As previously mentioned, this structure has been researched as a potential vapor sensor. Results of the initial research are shown in Figure 3.5 in which sections (a) and (b) show SEM

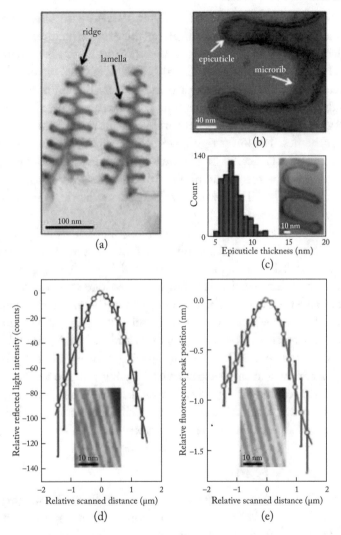

Figure 3.5: **Results of initial research to determine the polarity along a ridge of the Blue Morpho butterfly.**

images of the ridge structure with anatomy designated and the observed thickness of the outermost layer of material, the epicuticle. Use of a TEM shows the epicuticle has a thickness of approximately 7 nm. The results shown in sections (d) and (e) of the reflected light intensity and the fluorescence peak were used to determine the location of the ridge tops and the variation in fluorescent peak which aids in determining the polarity of different regions.

The results of Figure 3.5 (d) and (e) coupled with other experimental results show that there is a surface polarity gradient along the ridge structure. The tops of the ridges are more polar than the bottom of the ridges, as shown in Figure 3.6. Recall that many molecules and chemical substances are non-uniformly charged entities resulting in the variations in polarity along the ridge structure that may attract or repel different molecules. The change in reflectance as a result of absorbed vapors at various positions along the ridge is also shown.

Again, Figure 3.7 represents various structures and the simulation of different vapor deposition schemes. The lower right-hand picture shown in the figure (section (f)) is a clear representation of the criticality of the physical structure and how it relates to reflected color. The set of squares at the top of the photo are structures that consist of ridges with lamella and are shown to reflect blue. The lower set of squares are fabricated with only the ridge structure and are shown to reflect red.

In addition to this example of creating vapor sensors by replicating the structure of the butterfly wing, other potential applications include photonic security tags, protective clothing, cosmetics, and self-cleaning surfaces. By studying different species within the Morpho genus, researchers have learned that it is not only the ridge structure that contributes to the observed colors but also the upper and lower lamina that have a contribution.

The SEM images shown in the middle of Figure 3.8 ((i) through (l)) show a top-down view of the wing scale. The scale bars in (i) and (k) are 2 μm and in (j) and (l) 1 μm. For this particular Morph species, helenor, the spacing between the ridge rows are fairly large with the ridge rows narrow and smaller cross ribs. Spacing can also vary as, again, shown in the figure.

It is this compound complexity of structure, dimensions, and materials coupled with the aspect of the wavelength dependence on specific interactions such as reflections, refraction, critical angle, and absorption that results in these amazing color displays.

3.1.2   GENUS PAPILIO

The butterflies in genus Papilio fall into the "swallowtail" category. Most of the butterflies in this group that have the structural coloration live in the southern hemisphere whereas many of the North American swallowtail butterflies rely on pigment coloration. Hence, of interest for the nanoscale are the genus members from warmer climates.

The three butterflies of interest are the *P. blumei*, the green swallowtail, the *P. Ulysses*, the Ulysses butterfly, and *P. peranthus*, the blue swallowtail.

Members of the Papilio genus with structural coloration are shown in Figure 3.9. The word "Papilio" is Latin for butterfly. Sections (a), (b), and (c) of the figure are images of the *P. peranthus*, *P. blumie*, and *P. Ulysses*, respectively. These three images are at a normal viewing angel to the butterfly. Image (d) in the figure is the Ulysses swallowtail viewed at an angle. In this image the butterfly's wings appear blue rather than green. This is a true representation of the word "iridescent." Although normally implied to mean pearlescent or shiny, the definition of iridescent applies to any surface where the color observed is dependent upon the viewing

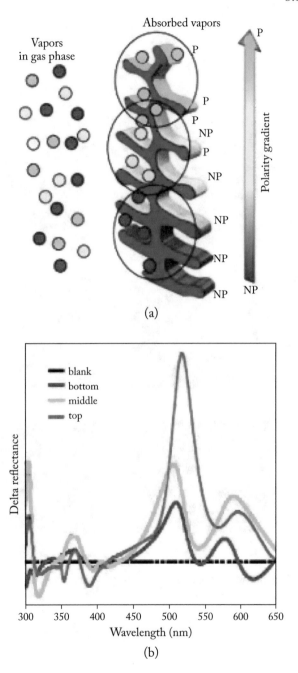

Figure 3.6: Gradient of surface polarity along the ridge structure and the change in the reflective response for various ridge positions.

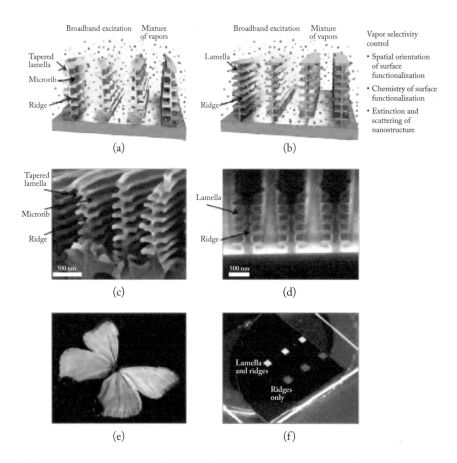

Figure 3.7: Simulation images of different physical structures of the ridge and lamella structure and vapor depositions, SEM images of the different structures, and results of optical tests.

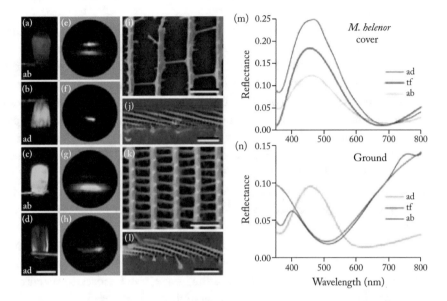

Figure 3.8: Images of the scales on the Morpho helenor species which shows the differences with the ridge and lamella structure.

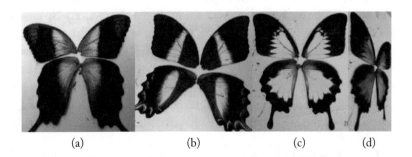

Figure 3.9: Butterflies in Papilio genus.

angle. The *P. Ulysses* butterfly is indeed iridescent, since when viewed from one angle it appears green and when viewed from a different angle it appears blue. Of note is the fact that in some nanoscale structures where structural coloration is observed iridescence is also observed, whereas in other situations it is not present.

SEM imaging of the scales on the *P. Ulysses* wing show that there are two kinds of scales present. Some of the scales are non-reflecting and have a serrated edge. Other scales with a smooth edge serve to reflect the light. The two types of scales are shown in Figure 3.10.

The scales that have the serrated edges do not reflect light and it was deduced that they constitute the dark regions on the wing. The observed colors in the different *Papilio* butterflies are due to different physical properties in the scales of each of the three types studied and it has been determined that all three of the *Papilio* species have a nanoscale layered structure.

The general structure of the butterfly's scales in the *Papilio* genus are similar to that of the Morpho genus and include longitudinal ridges with cross ribs and upper and lower lamina as shown in Figure 3.11. However, there are significant differences in the nanoscale structure that result in unique color reflections and angular dependence of the observed colors as exhibited in Figure 3.9.

Shown in Figure 3.12 is an image obtained using a Focused Ion Beam (FIB) machine to create a cross section of the scales, where the scale bar in the image is 1 μm. In this figure, which has the top of the scale at the bottom of the image, two nanoscale aspects can be observed. First, it is somewhat concave in nature and, second, it is composed of multiple layers (7). The layers are composed of a cuticle material alternating with an air-laminae. This air-laminae has pillars of cuticle material seen as the vertical bars that separate large air pockets. The resulting lamina complex has the cuticle laminae with an index of refraction around 1.56 and the air layer (with cuticle pillars) has an index of refraction close to that of air. Therefore, this structure acts as a Bragg reflector. Just as in the wing scale, a Bragg reflector, Figure 3.13, can consist of layers of materials, with different indices of refraction so that light is reflected and refracted at each interface. In the butterfly scales there are multiple layers of air-lamina and the cuticle structure which will reflect, refract, and transmit different wavelengths of light.

Cross sections of the scales on both the *P. blumei* and *P. peranthus* showed they have structures composed of seven layers, while *P. peranthus* has eight layers. The cross sections also provided data on the thickness of both the cuticle and the air-laminae layers.

A combination of experimental testing and measurements, coupled with computer simulation, allowed the *derivation* of the effective index of refraction for the two different types of layers.

Recall that the index of refraction is defined by the ratio of the speed of light in a vacuum to the velocity of the light in the material, and the velocity is dependent upon the wavelength of the light—which is the color. The angles which define reflection, refraction, and total internal reflection are dependent on the index of refraction of the material and the thickness of the material.

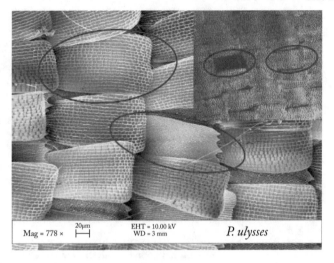

Figure 3.10: SEM image of the *P. Ulysses* wing.

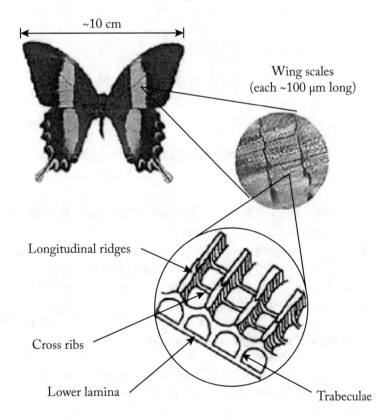

Figure 3.11: Detailed structure of the *Papilio* wing scale.

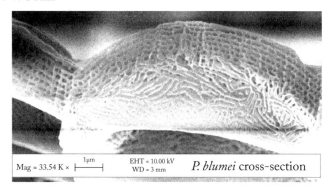

| Mag = 33.54 K × | 1μm | EHT = 10.00 kV
WD = 3 mm | *P. blumei* cross-section |

Figure 3.12: Cross-sectional view of the wing of *P. blumei*.

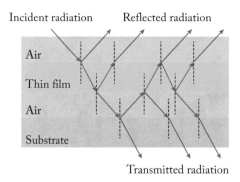

Figure 3.13: A general Bragg reflector with multiple different materials and interfaces.

This combination of factors and dependencies results in the brilliant colors as well as the angular dependence of the colors observed.

Table 3.1 shows that variations among three different species, each with the layered structure of cuticle laminae and air-laminae for several factors. Although the values for all the components are similar in thickness and index of refraction, there is still a significant difference in observed color, as shown in Figure 3.9.

The last column is the optical thickness, although it is not a true thickness. It represents the ratio of the incident radiant power to the transmitted radiate power. Power is related to energy, which is related to wavelength and index of refraction, which will determine observed color at different angles. As the optical thickness is reduced, transmitted power is increased, hence energy is increased. Blue wavelengths have a higher energy that green or red which draws the relationship between color and optical thickness. In general, as the optical thickness is reduced, wavelengths toward blue are reflected. In addition, the shorter periodicity of the *P. Ulysses* causes

Table 3.1: Summary of average thicknesses, index of refraction, and optical thickness

	Cuticle Laminae Thickness d_1	n_1	Air Laminae Thickness d_2	n_2	$n_1 d_1 + n_2 d_2$
P. peranthus	93 nm	1.56	108 nm	1.224	277 nm
P. blumei	102 nm	1.56	96 nm	1.168	271 nm
P. Ulysses	89 nm	1.56	100 nm	1.112	250 nm

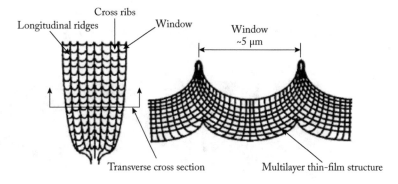

Figure 3.14: Cross-sectional drawing of the observed scale structure.

the color to change from a greenish-blue hue when viewed from the normal (perpendicular) angle, to bluer when viewed at an angle.

Even in early research [33], the unique structure of the wings of butterflies was acknowledged. Figures 3.14 and 3.15 represent the understanding of the micro-scale structure of the scales on the wing of the *P. blumei* butterfly and the underlying physics phenomena of thin film interference as a mechanism for the observed colors.

Note the concave structure in the drawing and the commonality with the concave structure shown in Figure 3.12 15 years later.

The concave structure adds an extra dimension to the coloration scheme of this particular genus of butterfly, and has been replicated based on analysis of the natural structure shown in Figure 3.16. Sections (a)–(c) are macro-scale images of the butterfly and the wing structure with the two different types of scales, reflective, and non-reflective.

The observation of this particular butterfly shows combinations of different colors in the yellow and green ranges. These different colors are reflected from different areas of the scales. The interactions are partially due to the multiple layers contributing to constructive and destructive interference of the various light wavelengths, but also due to the concave nature of the scale surface.

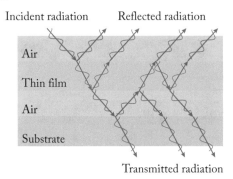

Figure 3.15: Light interactions in a thin film structure showing wavelength, reminder of the wavelength dependence.

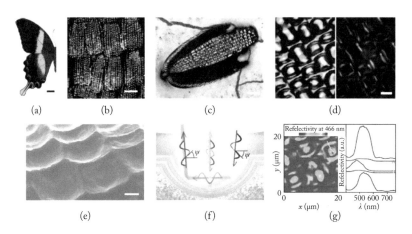

Figure 3.16: Results of analysis, testing, and observation of the wings of the *P. blumei* butterfly.

As previously mentioned, some of the wing scales appear to participate in the reflection of light while others do not. The non-reflective scales lay under the reflective scales and absorb transmitted light, thereby preventing it from being backscattered and jeopardizing the purity of the reflected light.

Section (e) of Figure 3.16 is an SEM image of the surface of the scale and shows the array of concave structures. The drawing in section (f) of Figure 3.16 shows how the multilayer aspect of the structure reflects a yellow-green combination of light from the center and blue from the edges, acknowledged as color mixing. The interaction of incoming light with this concave structure includes aspects of geometrical polarization, polarization conversion, and relative phase shift-induced ellipticity, and includes double and triple bounces of the incoming light before reflection. On each of these bounces, as the light interacts with the multiple layers of the concave

Figure 3.17: The Glasswing Butterfly, species *Greta oto* (anywhere.com/flora-fauna/invertebrates).

structure changes in the wavelength of the light reflected at a given angle can occur, changing the peak reflectance color.

These two sections provide examples of not only the multiplicity of nanoscale structures that can exist in nature but also the intricate ways that impinging light can be modified for the "viewer." The glasswing butterfly, Section 3.1.3, has a structure that allows for no color to be observed.

3.1.3   GLASSWING BUTTERFLY

Researchers have been intrigued by the photonic structures of butterfly wings which result in the reflected colors and the iridescence or angle dependence of the color observed for decades. Similarly, the transparency of the wings of the glasswing butterfly pose a puzzle.

The glasswing butterfly is found abundantly in Central America year-round. It is a moderately sized butterfly with a wingspan of 6 cm. The uniqueness of this insect is obvious in that most of its wing is transparent.

The wings of the butterflies shown in Figure 3.18 were tested for reflectance and transmittance of light of different wavelengths at different angles. The results of this testing are presented in Figure 3.19. Sections (a) and (b) show the results of reflectance vs. wavelength at different angles for the morpho and glasswing butterflies, respectively, with (c) showing a comparison of transmittance vs. wavelength. Note the vertical scale differences for (a) and (b). Comparison of these results shows that the Morpho has a peak reflectance percentage at all angles in the blue wavelength range whereas the Glasswing has minimal reflectance for all wavelengths at various angles. Section (c) shows a clear difference in the transmittance of the two. Sections (d) and

(a)   (b)

Figure 3.18: Photographs of Morpho and Greta genus butterflies.

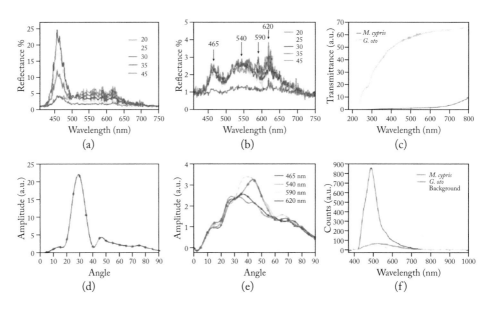

Figure 3.19: Results of reflectance and transmittance testing for the *M. cypris* and the *G. oto* butterflies.

(e) again compare the *cypris* and the *oto* for amplitude or reflectance intensity at different angles. Again, note the difference in the vertical axis scales. The *cypris* butterfly shows a clear high reflectance amplitude for 460 nm wavelength and 30° in section (d). By comparison, the *oto* exhibits low amplitudes for multiple wavelengths at various angles. Fluorescent counts, section (f), again express the difference in responses to light of the two species.

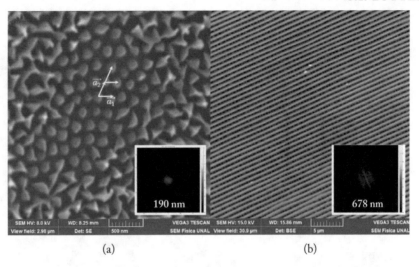

(a)   (b)

Figure 3.20: SEM images of the *M. cypris* and *G. oto* structures.

SEM imaging of the wing structure of the two butterflies shows a distinct difference. Figure 3.20 is the images with scale bars of 500 nm and 5 μm. The arrangement of "dots" observed for the *oto* were found to represent the nipple or chitin support structure for nanopillars of epicuticular wax (see Figure 3.21 (j)) and the *cypris* image shows the representative ridges as discussed in previous sections. The inset images are power spectrum density measurements.

Research performed in 2021 [21] provided more insight into the transparency of the *G. oto* species and revealed additional nanostructures, confirming earlier observations. The transparent regions of the wing contain a fewer number of scales than the non-transparent regions. There are also reflectance variations between scales with serrated edges and those with smooth edges. Transparent regions showed macro-scale hair-like structures and nanoscale structures including pillars of a wax-like material.

A compendium of results is shown in the images and graphs of Figure 3.21. The structures, attributes, and scale bar sizes are shown in Table 3.2.

The results shown in Figure 3.21 and summarized in Table 3.2 represent the continuous stream of new data and understanding about the transparent regions in the wings of the glass-wing butterfly. Both the images and data in Figure 3.21 show the structural aspects of the wing that lend to its transparent aspect.

Assessing and understanding the transparency contributors has led to consideration of applications involving sight and the structure of the eyeball. The creation of a better intraocular pressure (IOP) sensor is one such application. The IOP value is representative of a healthy eye or, with variations in the IOP values, an eye that may be diseased and leading toward blindness. By using biocompatible structures based on the nanostructures in the glasswing butterfly

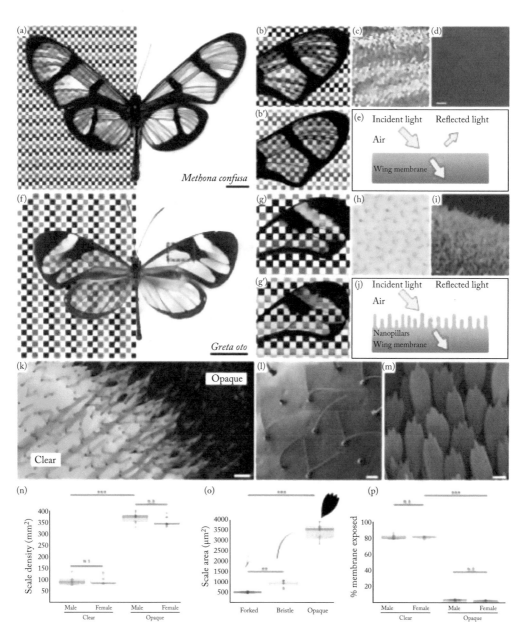

Figure 3.21: Results of multiple-stage investigation of the giant glasswing (*Methona confusa*) and the glasswing (*Greta oto*).

Table 3.2: Compilation of the data from Figure 3.21

Genus	Figure Section	Attribute	Scale Bar	Comment
Methona confuse	(a)	Butterfly	1 cm	
	(b)	Wing in reflected light		
	(b')	Wing with transmitted light		
	(c)	Clear wing portions showing reflection	100 μm	Reflection
	(d)	SEM of wing membrane	1 μm	Smooth, no nanostructures
	(e)	Diagram of reflection and refraction on the smooth membrane		Wing has a high refractive index, so light is reflected at the surface
Greta oto	(f)	Butterfly	1 cm	The red box shows the area to be studied and includes clear and opaque portions
	(g)	Wing in reflected light		
	(g')	Wing with transmitted light		
	(h)	High magnification of the clear region	100 μm	Minimal reflectance and bristle hairs can be seen
	(i)	SEM of wing portion	200 nm	Irregularity and nanoscale pillars resulting in omni-directional anti-reflectiveness
	(j)	Diagram of reflection and refraction on wing with nanostructures		Smoother gradient of the index of refraction between air and chitin
	(k)	Boundary between clear and opaques wing portions	100 μm	
	(l)	SEM of clear region	20 μm	Bristle like morphologies
	(m)	SEM of the opaque region		Large, flat-scale, serrated, typical pigmented morphologies
	(n)	Scale density measurements in clear and opaque regions		
	(o)	Surface area for different morphologies		
	(p)	Percent of wing exposed in the *G. oto* clear and opaque portions		

IOP sensors can be developed that are non-intrusive and provide more accurate and continuous readings. In this way a diagnosis can be made sooner, and preventative measures taken.

In summary, understanding the structure of the wings of some of nature's most beautiful butterflies has provided benefits not only because of the microscopes which have been developed that allow the observation and study of nanoscale structures but also by the complementary growth in semiconductor process advancement and material science understanding. These discoveries, and replications of the natural structures, have applications in the fields of paints and coatings, security, medical diagnostics and treatment approaches, sensor systems, and agriculture as well as applied sensors or coverings for early detection of structural failures.

3.2   BEETLES

3.2.1   NAMIB DESERT BEETLE

The Namib Desert located on the southwest coast of Africa is one of the oldest and also the driest deserts in the world. The desert environment is a combination of high winds and temperatures and dense fog in the early morning. Very few people, animals, plants, or insects can survive the severe environment. Nevertheless, several beetles of the genus *Stenocara* have been able to thrive there, survive this harsh environment, and even harvest water from the air. There are several species that exist in the desert and the species *Stenocara gracilipes* is most known as the Namib Desert Beetle. Figure 3.22 shows this beetle in the upper left-hand image. A curiosity about the beetle and how it survives in that environment has existed for decades, in particular, how the insect manages to collect enough water to live.

Researchers have discovered that the secret lies in the outer layer of the beetle's shell. The shell has regions that are superhydrophilic (water loving) and other areas that are superhydrophobic (water fearing). The beetle is able to collect water from the apparently dry air (discussed later) and from the early morning fog.

In the fog the water "droplets" are plentiful but very small with an average diameter of 10–15 μm. Because of the shape of the shell, water droplets that are collected on the top or upper regions of the shell, forming larger drops, will roll down into the beetle's mouth.

Some regions of the beetle's shell are covered with a wax coating are hydrophobic, while regions that are "wax free," as shown in the middle image of section (a), are hydrophilic.

Shown in the far-right image of section (a) of Figure 3.22 is an SEM image of the region in between the bumps on the fused wings (elytra) of the beetle. Therefore, the overall structure of the fused elytra shell of the beetle consists of bumps that are predominantly wax free. These bumps are randomly arrayed and are approximately 0.5–1.5 mm apart and about 0.5 mm in diameter. The bumps are superhydrophilic so water molecules or droplets will attach to them. The lower regions are superhydrophobic and composed of a textured surface shown in (b) of the figure.

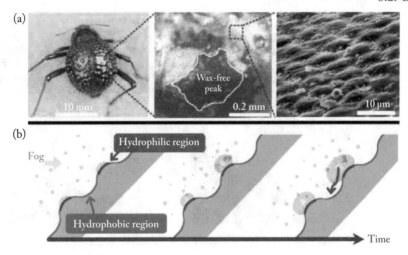

Figure 3.22: Images of the shell of the Namib Desert Beetle at different scales and the movement of a drop of water collected from a fog environment.

The lower portion of Figure 3.22 is a series of drawings showing the movement of a drop of water down the shell of the beetle. This drawing assumes a fog environment and as the water droplet moves over the surface water can be added to the drop.

The forces involved in this phenomenon include the cohesive force between the "like" molecules in the water droplet creating the surface tension that keeps the droplet "in shape," and the adhesive forces between the droplet and the surface. The strength of the adhesive force is dependent on the contact area and the surface chemistry. On the waxy superhydrophobic surface the adhesive force will be much smaller than that force of the superhydrophilic regions.

Because of the fundamental requirement of water for life, the study of water collection has occupied scientists for centuries. Different models have been developed which describe the measurements that are required and how they are obtained to understand the interactions of water with surfaces and how to compare different surfaces.

Drawings representing the different theories and water-collecting mechanisms are shown in Figure 3.23. The different approaches, equations, and dependencies are provided in Table 3.3.

The details, assumptions, and specific dependencies for each of the equations can be found from various sources such as [5].

Young's Equation from 1805 ignored any characteristics of the surface such as roughness, chemical heterogeneity, and so on. Acknowledging that surface roughness is a critical parameter, especially when dealing with water contact, Wenzel developed an equation to account for the modification to the "pure" contact angle due to roughness and is shown in the second line of the table. The parameter r is the ratio between the surface contact line or area (consider as the liquid covers the pillars in (b)) to the contact area for the liquid and air. For a smooth surface

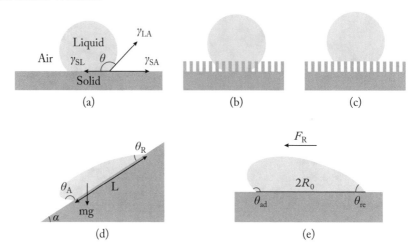

Figure 3.23: Drawings showing parameters for different wetting configurations between the liquid and the surface.

$r = 1$, for a rough surface $r > 1$ because the area of contact over the roughness will always be greater than on a smooth surface. Remembering that the interaction of a liquid with a surface is defined by the contact angle θ, this value will vary dependent on surface properties of roughness and chemical composition. When comparing surfaces of the same chemical composition, θ_0 represents the contact angle of the liquid on a smooth surface. The factor $(r \cos \theta_0)$ accounts for the total roughness of the surface.

There is a mathematical problem with the Wenzel model in that $r \cos \theta_0$ may be greater than 1, which creates an issue with the left side of the equation. Hence, the Cassie model was developed and is shown in the third line in Table 3.3. Clearly, the interaction of a liquid with any surface is a complex interaction dependent upon multiple constraints.

In the fog environment, water is collected in the superhydrophilic regions of the shell surface. Once the critical volume of a droplet has been reached, the force on the droplet due to gravity and the angle of the shell will overcome any adhesive force keeping the drop attached to the shell and the drop rolls toward the mouth of the beetle. Collection of water by the fog collection mechanism is based on the collision of the fog droplet and the superhydrophilic surface.

The external environment can also play a factor. In the case of the Namib Desert, the strong winds can positively interact with the beetle's shell resulting in the ability to collect water and keep the molecules and droplets captured.

This method of water collection is by means of dew condensation, which is a different process than fog collection. Dew is formed by the condensation of water vapor into droplets on a substrate. In order to create this water vapor, nucleation must occur. In other words, a defined amount of energy is required for the formation of a liquid nucleus on a flat surface. The energy

Table 3.3: Various theories for water collecting and surface interactions

Theory	Figure 3.23 Section	Equation	Variable Designation	Comments
Young's equation	(a)	$\cos \theta_0 = (\gamma_{SA} - \gamma_{SL})/\gamma_{LA}$	θ_0 is contact angle, γ is surface tension for SA, solid/air, SL, solid/liquid and LA, liquid/air	See Figure 3.23
Wenzel model	(b)	$\cos \theta_W = r \cos \theta_0$	θ_W is water contact angle, r is the ratio of the contact area water to surface (real) and the projected contact line	Include angle correction for surface roughness, values of $r \cos \theta_0$ may be >1 which causes math issues
Cassie model	(c)	$\cos \theta_C = r_f \cos \theta_0 + f - 1$	θ_C is the apparent contact angle, f is the solid water fraction under the contact area	Includes chemical heterogeneities, accounts for air pockets under the droplet
Furmidge equation	(d)	$mg \sin \alpha = \sigma \omega (\cos \theta_R - \cos \theta_A)$	m is weight, σ is surface tension, ω is contact circle width of the liquid droplet, α is sliding the angle, g is acceleration due to gravity, θ_A advancing contact angle, θ_R is the receding contact angle	For dynamic wettability evaluation of a liquid repellent surface
Sliding force	(e)	$F_R = \pi R_0 \gamma_{lv} (\cos \theta_{re} - \cos \theta_{ad})$	R_0 is the contact radius, θ_{re} receding contact angle, θ_{ad} advancing contact angle	Droplet assumes a hysteresis shape

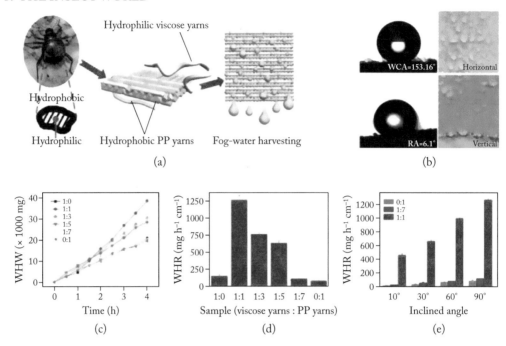

Figure 3.24: Using the Namib beetle shell structure as a template for a woven structure can be applied to harvest water.

required is dependent upon the wettability of the surface, hence, it is related to the contact angle. Wettability is a measure of a liquid to maintain contact with a surface. The wettability of a surface has a strong effect on the nucleation rate. The energy required for nucleation increases with contact angle. The larger the contact angle, the more hydrophobic the surface is, the lower the wettability value and the lower the nucleation rate. So that a hydrophobic surface which has a higher contact angle is therefore more difficult for nucleation or the gathering of water molecules to occur.

In this manner the beetle's shell surface, with regions of hydrophobicity and hydrophobicity, will capture water by vapor to liquid nucleation onto the hydrophilic regions. The droplets will grow by condensation, then by coalescing into large drops reach the volume where they roll down the sides of the shell and into the beetle's mouth.

Figure 3.24 shows how this capability has been replicated in a woven fabric as a water harvesting structure based on the Namib desert beetle. Testing of the materials shows a water contact angle of over 153° (section (b)) and a roll-off angle of 6.1°. The lower three graphs represent the growth of the structure by weight vs. time and the water harvesting rate vs. sample ratios and angle of inclination.

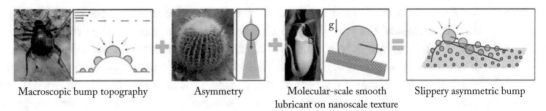

Macroscopic bump topography   Asymmetry   Molecular-scale smooth   Slippery asymmetric bump
                                             lubricant on nanoscale texture

Figure 3.25: Adding the understanding and replication of various aspects of nature can equal a solution for water collection, control and distribution.

Figure 3.26: The diabolical ironclad beetle.

Replication of the combination of superhydrophilic and hydrophobic materials coupled with an angled surface has the potential to provide a cost-effective, lightweight and energy-efficient method to capture and collect water from the air, even in extremely dry regions. When combined with lessons learned from other aspects of nature as exhibited in Figure 3.25, a solution to the problem of global thirst could be solved.

3.2.2   DIABOLICAL IRONCLAD BEETLE

The Diabolical Ironclad Beetle (*Phloeodes diabolicus*), Figure 3.26, is about 2 cm in length and is found in North America.

At first glance it does not appear to be much different from other desert and flying beetles and other jointed invertebrates. All have an arthropod exoskeleton which is a structure composed of four layers. Of most importance for the Ironclad beetle are the upper or outer segment, the epicuticle, and the procuticle which is composed of two layers, the exocuticle and the endocuticle. Molecules of polysaccharide α-chitin combine with proteins and form fibers, which are the primary structure in the cuticles. These fibers then form a multi-layered laminate structure which is tough, energy absorbent, and can twist. Not only that but the interfaces between the layers serve to inhibit crack propagation. The combination of these features are present in the many insects that are tolerant to high levels of pressure and force and can squeeze into small cracks and spaces.

However, the sum of these properties is not sufficient to explain the outstanding toughness of the diabolical ironclad beetle which can survive a car driving over it without a scratch and can withstand a force equivalent to a weight of 15 kg, or about 39,000 times its body weight. The question is: "What makes this beetle so strong?"

One unique aspect of this beetle is the fact that the forewings, the elytra, are locked together in a joint called a suture. This suture runs the length of the elytra and is not found in other beetles. With the forewings permanently joined together the ability to fly has been lost by the ironclad beetle. Nevertheless, the joining of the elytra and the standard exoskeleton structure still cannot explain the observed strength of this insect.

Using micro-computed tomography, the structure of the ironclad beetle has been investigated. One discovery was that at the interface between the elytra on the dorsal (top) side of the beetle and the ventral cuticle on the bottom of the beetle there existed lateral supports. It was also noted that there appeared to be three different types of lateral supports. The supports were named interdigitated, interlocking, or latching, and free standing. These features of the elytra and the ventral cuticle are shown in Figure 3.27. Also, each type of support was found in different segments of the abdomen of the beetle.

The segments of the abdomen where the different types of supports are found are shown in Figure 3.28. Interdigitated supports are found in the blue region, latching or interlocking supports are found in the magenta sections and the free-standing connectors in the green colored segment. Section (a) of the figure also shows a side view of the insect with the black region showing the open space for organs. The strongest support, interdigitated, is found in the segment of the insect with the densest concentration of organs to protect. Section (b) of the figure includes false-colored SEM images showing the interfaces between the elytra and cuticle for each of the support types. The lower-right image is the microtrichia, (small hairs) which cover the surface and are found in the latching support structure shown in magenta in section (a). The microtrichia provide friction to the support interface which reduces the ease of movement or sliding. Section (c) is a graph which expresses the relative strength of each of the supports. The interdigitated support is listed as the first support and is the strongest of the three. Section (d) presents computer simulations of the three different support types and the stress distribution.

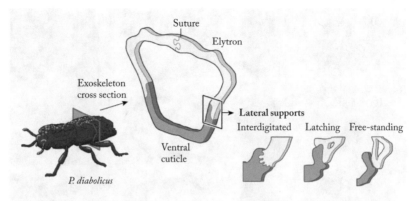

Figure 3.27: The ironclad beetle with a cross section shown with location of the suture and lateral support structures. As shown in the diagram each type of lateral support has a different configuration for the interlocking of the elytra (light gray) and the ventral cuticle (dark gray).

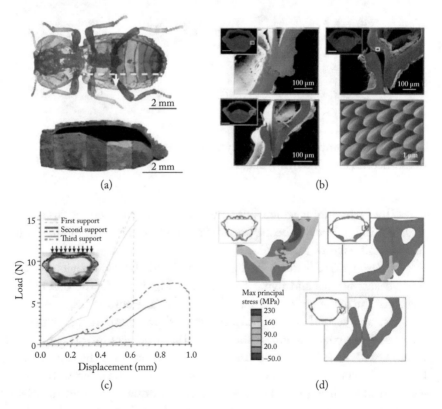

Figure 3.28: Color-coded division of the beetle abdomen representing the presence of different connectors and additional detail.

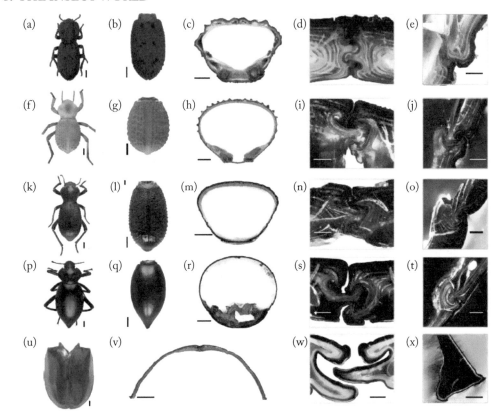

Figure 3.29: Comparison of the elytra suture structure and lateral connect structures for four desert beetles and one flying beetle.

The suture joint which joins the two sides of the elytra is, in itself, a unique bonding structure in the ironclad beetle and looks similar to the interdigitated interface and mimics interconnecting jigsaw puzzle pieces.

These unique features of the ironclad beetle become obvious when it is compared with other desert beetles and flying beetles as in Figure 3.29. The desert beetles are the top four insects in the figure with the ironclad at the top. The bottom beetle is one with flying capability. The first and second columns of the figure show photographs of the beetles and the elytra. Column 3 is a cross section of the elytra. The scale bars in this column are 1 mm. The fourth column, with scale bars of 100 μm, shows the structure of the suture joining the elytra. As can be seen in column 4, the interconnecting structure of the ironclad beetle is continuous over the entire meeting distance and is dissimilar to the other elytra and especially to the flying beetle where the elytra are not in physical contact at all. The last column shows lateral support examples for the various species, again the scale bars are 100 μm.

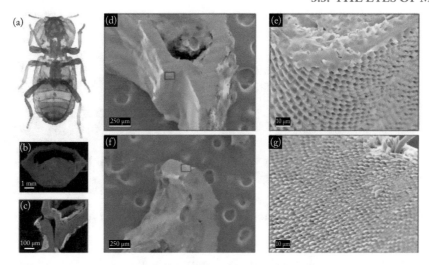

Figure 3.30: CT scan of the ironclad beetle (a), (b), and (c), and SEM images of the microtrichia.

The presented analysis of the physical structural variations of the diabolical ironclad beetle substantiates the enhanced strength attributes of the beetle. SEM images of the frictional microtrichia on the interface surfaces are shown in Figure 3.30.

Sections (a), (b), and (c) of the figure are CT scans of the insect. Section (a) shows the color-coded regions of different lateral support. Section (b) represents a cross section of the support with the area in the red rectangle magnified in (c). The SEM images (d) and (f) are the magenta and blue regions of section (c), respectively. {Note the shape of the structures in each figure section.} Section (e) is a magnification of the red box region shown in section (d). Similarly, section (g) is a magnification of the blue box region shown in section (f). Both (e) and (g) show the microtrichia structure that lends cohesiveness and strength to the free-standing support.

Through multiple research investigations into the initially observed strength and crush-proof capability of the diabolical ironclad beetle, an understanding of the unique physical design and architecture of this beetle is being discovered. This understanding will support the development of large construction and minute medical applications.

3.3   THE EYES OF MOTHS

Moths are known as night fliers, with most species active only at night. As a result, they have several features, such as their coloration, that allow them to seek food and mates at night while remaining "hidden" from predators. One of these features is non-reflecting eyes. That is, the eyes of a moth absorb light of multiple wavelengths rather than reflecting that light. This feature has intrigued entomologists for decades as research has sought the explanation.

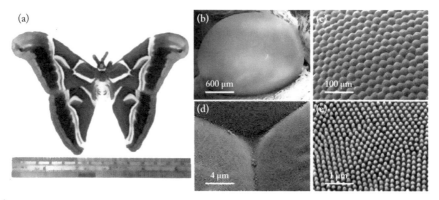

Figure 3.31: The silkmoth moth and SEM images of the eye structure.

As early as 1967, researcher C. G. Bernhard proposed that the array of conical protuberances on a moth's eye had a graded index of refraction in the vertical direction.

A graded index of refraction allows flat lenses to be used in place of curved or spherical lenses, reducing space allocations and allowing for multiple lenses to be placed next to each other for copiers and scanners. The lens in the human eye has a graded index of refraction being higher in the central portion and reducing at the outer portions. This provides a reduction in aberrations and distortion for both short and long distances. The reason that the sun can be seen even after it has "set" is because the Earth's atmosphere is a material with a graded index of refraction.

Figure 3.31 includes a photograph of a silkmoth, *Philosamia cynthia ricini*, and SEM images of the moth's eye at various magnifications. Use of various approaches for creating anti-reflective coatings and structures is not a recent development. The standard approach has been layered thin films which result in reduced reflection because of destructive interference in the reflected waves. Although a successful approach, a given structure is only applicable for a small range of wavelengths. Vapor deposition, a process used in the manufacture of semiconductor devices, can be used to create more complex structures with precise thicknesses unlike the first process. These created thin films create a layer of material with a graded refraction index between the air and the glass substrate.

The eye construction of the moth contains structures approximately 250 nm high in a neat array with a periodicity of 300 nm, as shown in Figure 3.31. Note that visible light is in the 400–700 nm range, therefore the structures in the eye are in the sub-visible wavelength range. Light interactions such as reflection and refraction are dependent upon the relationship between the wavelength of the light and the index of refraction of the material. In addition, any time there is an abrupt change in the materials, which occurs when the materials are in layers, an interaction with the light will occur.

Because the structure of the moth's eye contains features at the sub-visible wavelength and because there are no abrupt changes in material impacting light interaction, the structures act

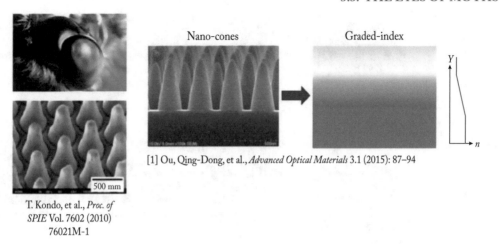

[1] Ou, Qing-Dong, et al., *Advanced Optical Materials* 3.1 (2015): 87–94

T. Kondo, et al., *Proc. of
SPIE* Vol. 7602 (2010)
76021M-1

Figure 3.32: A moth eye and SEM images of the structures and the resulting graded index of refraction.

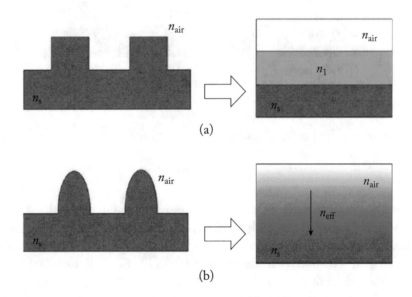

Figure 3.33: Subwavelength structures resulting in different profiles for the index of refraction.

as a material with a graded index of refraction. The result of this combination of effects is that the light is not reflected toward the outer surface. This is represented in Figures 3.32 and 3.33.

Both structures represented in Figure 3.33 have dimensions that are smaller than the wavelength range of visible light. In the (a) portion of the figure the rigid structures will result in an effective graded refractive index yet still have distinct boundaries governed by the ratio between

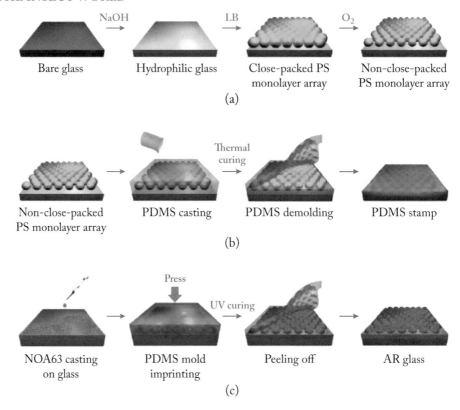

Figure 3.34: The proposed molding process to replicate the anti-reflective properties of moth eyes.

the ridges and channels. In the (b) portion of the figure where the structures are rounded and parabolically shaped the graded refractive index is a gradual change from one index of refraction to the next. The resulting refractive index depends on the amount of material per layer and is based on the effective medium theory (EMT). EMT is a mathematical method to average multiple index values of various constituents that make up a composite material.

As previously mentioned, there have been several methods applied to the replication of the moth eye and creation of anti-reflective materials and coatings. The first method consists of merely spraying or painting the layers using templates to create the structures. The second process uses a semiconductor fabrication process to apply very defined layers.

Another method is being evaluated which is shown in Figure 3.34. This approach uses readily available polymer-based materials and chemicals in a deposition and molding process. This process has the potential to be more accurate than the first approach and more cost effective than the second approach. Finally, the third proposed approach, could be used to create structures

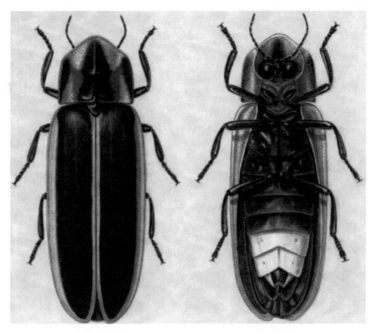

Figure 3.35: A drawing of the dorsal (top/front) side and the ventral (bottom) side of a firefly.

with a large surface area, which is a challenge especially for the semiconductor fabrication-based approach.

New process approaches are continuously being developed [7] which address challenges related to use of various substrate materials, more available materials, and aspects of cleaning and contamination which can reduce the applicability and lifetime of some coatings.

Anti-reflection is a desired property for many applications from screens to light-emitting diodes and sensor systems. By continuing to study and replicate the structure found in the eyes of moths, efficient and cost-effective solutions are being found.

3.4   FIREFLIES

Fireflies have intrigued children, adults, and researchers for decades—as have most aspects of the natural world around us. Recent research has discovered that the firefly's "light" or luminosity and the large distance over which it can be seen are a result of three nanotechnology-related aspects of the firefly's lantern.

Figure 3.35 is a drawing of the top and underside of the firefly beetle. The lantern, which is the light-emitting portion of the insect, is clearly visible as the yellow segments of the firefly abdomen. It is in this region that the three contributing factors to the firefly's light are found.

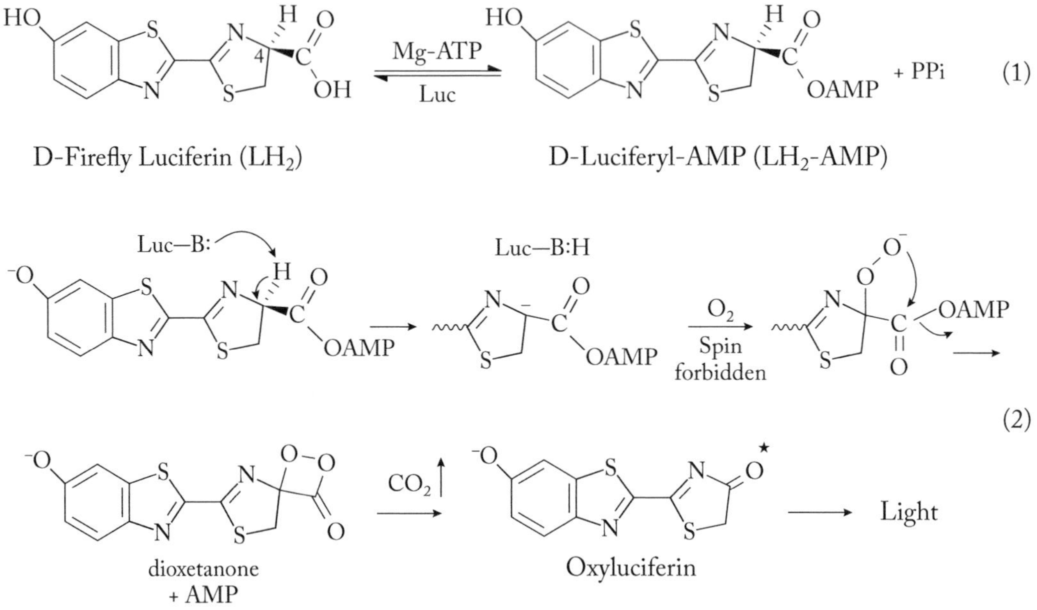

Figure 3.36: Two Luc-catalyzed half-reactions that result in the firefly bioluminescence.

First, the luminescence is caused by a biologically based chemical reaction, hence the name bioluminescence. In this chemical reaction, luciferin, an organic molecule interacts with an energy source, adenosine triphosphate (ATP), an enzyme and oxygen to create a chemical called oxyluciferin along with a few other biproducts. The resulting oxyluciferin molecule has an electron that, through the chemical reaction, has an energy higher than the electron energy in the ground or unexcited state. Electrons remain in this higher energy state for a very short period of time and quickly move down to the ground or lower energy state. In this process of transition, energy is lost, and that energy takes the form of visible light. This process is called single-electron transfer oxidation. The process is shown in Figure 3.36.

A unique aspect in the creation of this bioluminescence is the fact that the chemical reaction is organic and does not require coenzyme A, metal ions, or other coenzymes which are normally associated with reactions of this type. Applications which require bio compatibility or minimal chemical intrusion such as drug testing or monitoring water for contamination could benefit from the organic nature of the process and relatively simple interaction required to create the luminescence.

This light produced has a wavelength range of 552–582 nm [27]. The range of wavelengths is dependent upon several factors of the chemical reaction as well as temperature, and the wavelength varies among the different firefly species. It is this subtle wavelength variation that allows species to detect each other and attract mates of the same species.

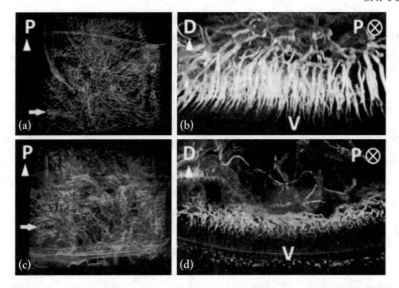

Figure 3.37: **Still shots from x-ray movies of the tracheal system of two firefly species.**

The process defined in Figure 3.36 which is responsible for the luminescence observed requires a supply of oxygen for the reaction to take place. Research to define and determine source of the supply of the required oxygen has involved the tracheal system of the firefly. The tracheal system in insects allows oxygen and carbon dioxide to travel through air-filled tubes. These tubes become much smaller when they near the ends or within tissue. The narrow tubes are called tracheoles. The tracheoles can be less than 300 nm in diameter. This research [36] has resulted in x-ray movies of the interactions and movement of the tracheoles. Still images from the movies are in Figure 3.37 and show the intricate structure of the tracheal system from two firefly species. The D and V represent the dorsal and ventral side of the insect, respectively, and P represents the posterior side.

By studying the movement of the tracheoles and measuring oxygen content and consumption within systems of the firefly they have determined that to supply enough oxygen to the light creating chemical process, the amount of oxygen to the mitochondria is reduced. The mitochondrial activity and therefore energy required, i.e., oxygen, is reduced by introducing nitric oxide (NO). NO also stimulates luminescence. So essentially, when luminescence is required the firefly's tracheal system ensures that enough oxygen is provided to the chemical reaction to generate the light while minimizing the oxygen required by other tissues. This shows an amazing and efficient use of energy and oxygen; the symmorphosis hypothesis (Taylor and Weibel) that biological structures meet the maximum functional requirements with minimum excess.

The second contributing factor to the firefly's luminescence is a result of the nanoscale structure found in the abdominal segments. Figure 3.38 represents many aspects of the nanoscale structure.

First of note is that the surface of the normal (N) segment on the dorsal side of the firefly is covered with many hairs (c), whereas the lantern segments, labeled L in section (b) have a minimum number of setae (d) and show a distinctly different surface structure than the normal segments. Note that the scale bars in both (c) and (d) sections are the same. Sections (e) and (f), of the figure are SEM images of the lantern segment and show not only the scale microstructures but also the longitudinal nanoscale structures. Section (g) is a Focused Ion Beam-(FIB) created cross section showing another aspect of the segment. That aspect is the inclination of the cuticle scales. Section (h) shows the high levels of transmittance from the structures. Sections (i) and (j) show drawn aspects of the structure. Finally, the lower portions of the figure provide drawings of the natural and manmade structure.

Continued research has focused on lessons learned by studying the lantern structure of the firefly. Work [15], focused on the finer nanoscale structure, specifically the cuticle of the lantern. The research focused on structures that could be used to enhance the light extraction for an energy efficient LED. Figure 3.39 presents the obtained SEM images and the evolution of a process that can be used to create stripes or nanoparticles.

Note that in section d of Figure 3.39 the scale bar on the image of the ridges is 550 nm. This matches well with the drawing of the bumps in Figure 3.38 (j). There are significant layers of nanostructures in the lantern outward portion of the firefly that will support extraction of the light generated by the internal chemical process.

The third aspect that contributes to the brightness observed in the firefly's light is the reflective coating that is at the back side of the lantern segment.

Viewing a cross section of the lantern segment, as shown in Figure 3.40 (c) and (d), the layers top to bottom of the cuticle, photogenic layer (PL) and reflective layer (RL), can be seen. The first light-generating and transmitting aspect of the firefly discussed was the chemical reaction that occurs in the photogenic layer. The second aspect was the nano structures comprising the cuticle of the lantern segment which serve to extract and distribute the light produced. This third aspect ensures that light which may originally be directed toward the "back" of the lantern segment gets reflected to exit out the front. Understanding this reflective structure will support the development of various photonic materials, sensitive detection, and imaging systems and medical diagnostics and therapy.

As shown in Figure 3.40 (f), the reflective layer is composed of small spheres with an average diameter of 1.12 μm. Using a transmission electron microscope (TEM) the spheres were determined to be hollow with a high reflectance in the visible range.

The RL has a high reflectance in the wavelength range of 400–800 nm and a reflectance of 82% around 550 nm which is the wavelength window of the light emitted by the PL (Fig-

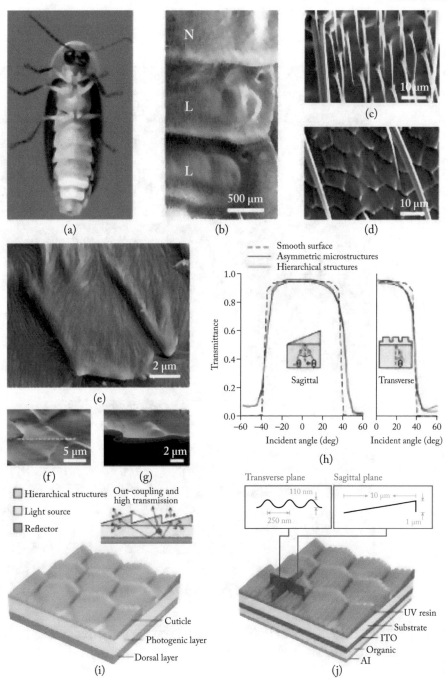

Figure 3.38: The firefly, abdominal sections, SEM images of the surface, transmittance data, physical structure, and man-made replication.

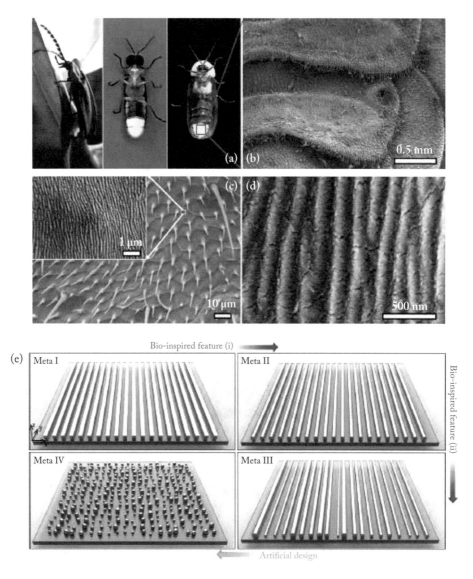

Figure 3.39: **SEM** images of the ridge structure on the lantern cuticle and process steps to create LED lens structure and nanoparticles.

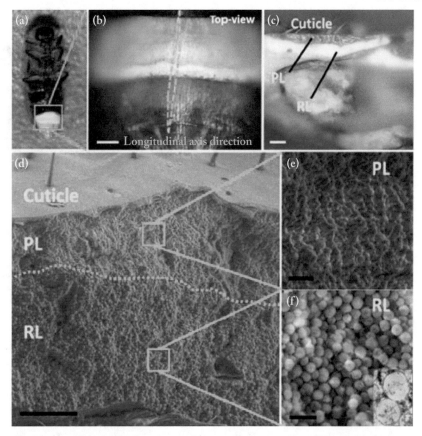

Figure 3.40: Optical, SEM, and TEM images of the layers within the firefly lantern segment.

ure 3.41). Therefore, the hollow spheres in the reflective layer effectively enhance the overall light intensity emitted from the lantern segment.

Nanotechnology and nanoscale structures play a critical role in the generation of the firefly's light and also in ensuring that the light can be seen for a very great distance. The understanding of the nanoscale structures will support development and applications from the medical field to communication.

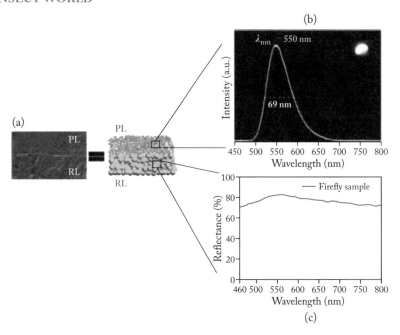

Figure 3.41: Graphical test results showing the predominant wavelength of light emitted in the PL and reflectance vs. wavelength in the RL.

3.5   REFERENCES

[1] Barrera-Patiño, C. P., Vollet-Filho, J. D., Teixeira-Rosa, R. G. et al. Photonic effects in natural nanostructures on Morpho cypris and Greta oto butterfly wings. *Scientific Reports*, 10:5786, 2020. https://doi.org/10.1038/s41598--020-62770-w DOI: 10.1038/s41598-020-62770-w.

[2] Branchini, B. R., Behney, C. E., Southworth, T. L., Fontaine, D. M., Gulick, A. M., Vinyard, D. J., and Brudvig, G. W. Experimental support for a single electron-transfer oxidation mechanism in firefly bioluminescence. *Journal of the American Chemical Society*, 137(24):7592–7595, 2015. DOI: 10.1021/jacs.5b03820.

[3] Chen, L., Shi, X., Li, M. et al. Bioinspired photonic structures by the reflector layer of firefly lantern for highly efficient chemiluminescence. *Scientific Reports*, 5:12965, 2015. https://doi.org/10.1038/srep12965 DOI: 10.1038/srep12965.

[4] Chen, P. Y. Tough lessons from diabolical beetles. *Nature*, 586:502–504, October 22, 2020. https://media.nature.com/original/magazine-assets/d41586--020-02840-1/d41586-020-02840-1.pdf DOI: 10.1038/d41586-020-02840-1.

[5] Chen, Z. and Zhang, Z. Recent progress in beetle-inspired superhydrophilic-superhydrophobic micropatterned water-collection materials. *Water Science and Technology*, 82(2):207–226, July 15, 2020. https://doi.org/10.2166/wst.2020.238 DOI: 10.2166/wst.2020.238. 61

[6] Clapham, P. and Hutley, M. Reduction of lens reflexion by the moth-eye, principle. *Nature*, 244:281–282, 1973. https://doi.org/10.1038/244281a0 DOI: 10.1038/244281a0.

[7] Diao, Z., Hirte, J., Chen, W., and Spatz, J. P. Inverse moth-eye nanostructures with enhanced antireflection and contamination resistance. *ACS Omega*, 2(8):5012–5018, 2017. DOI: 10.1021/acsomega.7b01001. 73

[8] Giraldo, M. A. and Stavenga, D. G. Brilliant iridescence of Morpho butterfly wing scales is due to both a thin film lower lamina and a multilayered upper lamina. *Journal of the Comparative Physiology A*, 202:381–388, 2016. https://doi.org/10.1007/s00359--016-1084-1 DOI: 10.1007/s00359-016-1084-1.

[9] Kim, J. J., Lee, J., Yang, S. P., Kim, H., Kweon, H. S., Yoo, S. H., and Jeong, K. H. Biologically inspired organic light-emitting diodes. *Nano Letters*, 16, 2016. DOI: 10.1021/acs.nanolett.5b05183.s001.

[10] Kim, J., Lee, Y., Kim, H. G., Choi, K. J., Kweon, H. S., Park, S., and Jeong, K. H. Biologically inspired LED lens from cuticular nanostructures of firefly lantern. *National Academy of Sciences*, 109(46):18674–18678, 2012. https://www.pnas.org/content/109/46/18674 DOI: 10.1073/pnas.1213331109.

[11] Kittle, J., Fisher, B., Kunselman, C., Morey, A., and Abel, A. Vapor selectivity of a natural photonic crystal to binary and tertiary mixtures containing chemical warfare agent simulants. *Sensors*, 20:157, 2019. DOI: 10.3390/s20010157.

[12] Kolle, M., Salgard-Cunha, P., Scherer, M. et al. Mimicking the colourful wing scale structure of the Papilio blumei butterfly. *Nature Nanotechnology*, 5:511–515, 2010. https://doi.org/10.1038/nnano.2010.101 DOI: 10.1038/nnano.2010.101.

[13] Kondo, T., Suzuki, A., Teramae, F., Kitano, T., Kaneko, Y., Kawai, R., Teshima, K., Maeda, S., Kamiyama, S., Iwaya, M., Amano, H., and Akasaki, I. Enhancement of light extraction efficiency of blue-light-emitting diodes by moth-eye structure. *Proceedings*, 7602, March 12, 2010. https://doi.org/10.1117/12.841518 DOI: 10.1117/12.841518.

[14] Kuo, W. K., Hsu, J. J., Nien, C. K., and Yu, H. H. Moth-eye-inspired biophotonic surfaces with antireflective and hydrophobic characteristics. *ACS Applied Materials and Interfaces*, 8(46):32021–32030, 2016. DOI: 10.1021/acsami.6b10960.

[15] Mao, P., Liu, C., Li, X. et al. Single-step-fabricated disordered metasurfaces for enhanced light extraction from LEDs. *Light: Science and Applications*, 10:180, 2021. https://doi.org/10.1038/s41377--021-00621-7 DOI: 10.1038/s41377-021-00621-7. 76

[16] McGuffey, M., Roman, S., Bittner, L., Griboff, Y., Wessler, G., Georgieva, M., and Osmond, J. Infrared antireflection moth's eye nanostructures. https://mse.umd.edu/sites/mse.umd.edu/files/IR%20Antireflection%20Moth%E2%80%99s%20Eye%20Nanostructures-2.docx.pdf

[17] Narasimhan, V., Siddique, R. H., Lee, J. O. et al. Multifunctional biophotonic nanostructures inspired by the longtail glasswing butterfly for medical devices. *Nature Nanotechnology*, 13:512–519, 2018. https://doi.org/10.1038/s41565--018-0111-5 DOI: 10.1038/s41565-018-0111-5.

[18] Ou, Q.-D., Zhou, L., Li, Y.-Q., Chen, J.-D., Li, C., Shen, S., and Tang, J.-X. Simultaneously enhancing color spatial uniformity and operational stability with deterministic quasi-periodic nanocone arrays for tandem organic light-emitting diodes. *Advanced Optical Materials*, 3:87–94, 2015. https://doi.org/10.1002/adom.201400337 DOI: 10.1002/adom.201400337.

[19] Park, K. C., Kim, P., Grinthal, A. et al. Condensation on slippery asymmetric bumps. *Nature*, 531:78–82, 2016. https://doi.org/10.1038/nature16956 DOI: 10.1038/nature16956.

[20] Parker, A. and Lawrence, C. Water capture by a desert beetle. *Nature*, 414:33–34, 2001. https://doi.org/10.1038/35102108 DOI: 10.1038/35102108.

[21] Pomerantz, A. F., Siddique, R. H., Cash, E. I., Kishi, Y., Pinna, C., Hammar, K., Gomez, D., Elias, M., and Patel, N. H. Developmental, cellular, and biochemical basis of transparency in the glasswing butterfly Greta oto. *bioRxiv*, 2020. https://doi.org/10.1101/2020.07.02.183590 DOI: 10.1101/2020.07.02.183590. 57

[22] Potyrailo, R., Bonam, R., Hartley, J. et al. Towards outperforming conventional sensor arrays with fabricated individual photonic vapour sensors inspired by Morpho butterflies. *Nature Communications*, 6:7959, 2015. https://doi.org/10.1038/ncomms8959 DOI: 10.1038/ncomms8959.

[23] Potyrailo, R., Starkey, T., Vukusic, P., Ghiradella, H., Vasudev, M., Bunning, T., Naik, R., Tang, Z., Larsen, M., Deng, T., Zhong, S., Palacios, M., Grande, J., Zorn, G., Goddard, G., and Zalubovsky, S. Discovery of the surface polarity gradient on iridescent Morpho butterfly scales reveals a mechanism of their selective vapor response. *Proc. of the National Academy of Sciences of the United States of America*, 110, 2013. DOI: 10.1073/pnas.1311196110.

[24] Provonsha, A. In appreciation of fireflies. *Home and Garden—Education Center*, Purdue Department of Entomology, Environment, 2011. https://www.purdue.edu/uns/html4ever/1998/980626.Turpin.fireflies.html

[25] Rivera, J., Hosseini, M. S., Restrepo, D. et al. Toughening mechanisms of the elytra of the diabolical ironclad beetle. *Nature*, 586:543–548, 2020. https://doi.org/10.1038/s41586--020-2813-8 DOI: 10.1038/s41586-020-2813-8.

[26] Sealy, C. Moth eye inspires antireflective coating. *Materials Today*, November 26, 2020.   https://www.materialstoday.com/nanomaterials/news/moth-eye-inspires-antireflective-coating/

[27] Seliger, H. H. and McElroy, W. D. The colors of firefly bioluminescence: Enzyme configuration and species specificity. *Proc. of the National Academy of Sciences of the United States of America*, 52(1):75–81, 1964. http://www.jstor.org/stable/72084 DOI: 10.1073/pnas.52.1.75. 74

[28] Siddique, R. H., Diewald, S., Leuthold, J., and Hölscher, H. Theoretical and experimental analysis of the structural pattern responsible for the iridescence of Morpho butterflies. *Optics Express*, 21:14351–14361, 2013. DOI: 10.1364/oe.21.014351.

[29] Song, B., Johansen, V., Sigmund, O. et al. Reproducing the hierarchy of disorder for Morpho-inspired, broad-angle color reflection. *Scientific Reports*, 7:46023, 2017. https://doi.org/10.1038/srep46023 DOI: 10.1038/srep46023.

[30] Stenocara gracilipes. *Wikipedia*, Wikimedia Foundation, April 12, 2021. https://en.wikipedia.org/wiki/Stenocara_gracilipes

[31] Strieter, A. Photograph of the Glasswing butterfly, species Greta oto. *Anywhere*, October 8, 2021. https://www.anywhere.com/img-a/eco/79/borboleta-asa-de-vidro-pereque-080621-pr121658a.jpg?h=300&fit=min&q=80

[32] Sun, J., Wang, X., Wu, J. et al. Biomimetic Moth-eye nanofabrication: Enhanced antireflection with superior self-cleaning characteristic. *Scientific Reports*, 8:5438, 2018. https://doi.org/10.1038/s41598--018-23771-y DOI: 10.1038/s41598-018-23771-y.

[33] Tada, H., Mann, S. E., Miaoulis, I. N., and Wong, P. Y. Effects of a butterfly scale microstructure on the iridescent color observed at different angles. *Optics Express*, 5:87–92, 1999. DOI: 10.1364/oe.5.000087. 53

[34] Tam, H. L., Cheah, K. W., Goh, D. T. P., and Goh, J. K. L. Iridescence and nanostructure differences in Papilio butterflies. *Optical Materials Express*, 3:1087–1092, 2013. DOI: 10.1364/ome.3.001087.

[35] Temmming, M. The diabolical ironclad beetle can survive getting run over by a car. Here's how. *Science News*, 198(9), 2020. https://www.sciencenews.org/article/diabolical-ironclad-beetle-exoskeleton-armor-impossible-squish

[36] Tsai, Y. L., Li, C. W., Hong, T. M., Ho, J. Z., Yang, E. C., Wu, W. Y., Margaritondo, G., Hsu, S. T., Ong, E. B. L., and Hwu, Y. Firefly light flashing: Oxygen supply mechanism. *Physical Review Letters*, 113:258103, December 17, 2014. https://doi.org/10.1103/PhysRevLett.113.258103 DOI: 10.1103/physrevlett.113.258103. 75

CHAPTER 4

From Plants

The purity of the lotus leaf and the "stickiness" of the pitcher plant have intrigued humans for centuries. Researching and understanding plants at the nanoscale has provided insight into these observations and applications in the modern world. Structures that exist naturally in some plants such as conifers can serve as natural water purification systems. Lessons from the plant world are only beginning to be understood.

4.1 LOTUS LEAVES

For centuries, the lotus leaf has been a sign of purity, predominantly because the leaves remain free of debris and dirt. Upon observation, a drop of water on top of a lotus leaf appears almost spherical, as shown in Figure 4.1. The water responds to the lotus leaf in this manner for several reasons.

First, recall that in any given situation, there are multiple forces at work. There are the atomic level forces, between individual atoms such as ionic and/or covalent bonding. There are also forces interacting between the individual molecules. In the case of water, a dipole molecule, the interaction between the water molecules results in a cohesive force. It is this cohesive force that holds the water molecules together in a "drop" and results in surface tension. Molecular cohesive force is the force between like (similar) molecules as in the drop of water. Correspondingly, when the molecules interacting are unlike each other (different types), such as between the water in a glass and the glass that holds it, or here between the water droplet and the surface of the lotus leaf, the resulting force is called an adhesive force.

From the observation of the drop of water on the leaf, it can be stated that the cohesive forces within the molecules of the water drop are much stronger than the adhesive forces between the water drop and the lotus leaf. That explanation, which is true, does not adequately explain the observed results or delve into the contributing factors to each of these interacting influences.

Hans Ensikat and co-authors [5] stated that the observed effect of a water droplet on a lotus leaf had to be due to "unrivaled optimizations." They attribute the amazing level of superhydrophobicity exhibited by a lotus leaf to a combination of the wax coating, micro and nano structures, and mechanical properties. Superhydrophobicity is the effect of adhesive interactions that result in the water droplet remaining almost spherical on top of the surface. In addition, it is not just the "wax" coating but the *structure* of the wax coating.

These higher-level features are shown in Figure 4.2. The top portion (a) is a photo of a lotus leaf. Portion (b) of the figure is an SEM image of the surface showing that the epidermis (outer

Figure 4.1: Water droplets on a lotus leaf in Florida wetlands. (Photo courtesy of Ryan Kennelly.)

Figure 4.2: A lotus leaf and SEM images of the surface showing papillae and wax tubules.

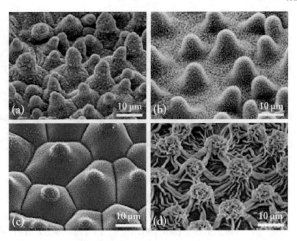

Figure 4.3: SEM images of papillae on various plants which exhibit superhydrophobic properties.

tissue) layer of the leaf is not flat but is covered by wax tubules. Section (c) in the figure shows the wax tubules. Note that (c) is a 10X magnification of (b). Image (d) is the leaf surface with the wax tubules dissolved and shows the stomata, or underlying structure, which are missing from the underside of the lotus leaf (e).

One of the common factors contributing to a hydrophobic surface is that the surface is not smooth and includes bumps of specific sizes and density on the surface. For a superhydrophobic surface, where the contact angle exceeds 140°, the bumps, protrusions, or in the case of plants, papillae, are very small and densely packed but not touching. Although there are indeed other plants that demonstrate superhydrophobic properties that have papillae covering the surface with a wax coating, and even some that have a smoother wax coating only, none match the superhydrophobicity of the lotus leaf.

A comparison of different plant surfaces is shown in the SEM images of Figure 4.3. Section (a) of the figure is the lotus leaf surface and the other three sections are images from other plants demonstrating superhydrophobicity. Note that the scale bars are the same for all four images so differences in the number per unit area and size of the papillae can be compared.

Figure 4.4 shows the measurements of the top surface area of the lotus leaf papillae as well as a cross section created by freeze drying the sample. The overall strength and resulting interaction between two non-similar materials is dependent on the contact surface area. In addition, the strength of the molecular cohesive bond in the similar-type molecule material (the water droplet), and the chemistry of the surface it is resting upon (the papilla surface). That is, the contact area, strength of the cohesive force and the chemistry of the surface all contribute to the strength of the adhesive force between two dissimilar materials.

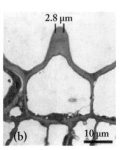

Figure 4.4: Lotus papillae showing tip dimensions and a cross section.

Even though the number of papillae per unit area is higher for the lotus leaf the area at the tip of each papilla is small and each tip area is not identical. This variation in papilla height and area will also contribute to the interaction of water with the surface.

Minimizing the contact area between the lotus leaf papillae and the water drop will reduce the adhesive force between the non-similar materials and reduce the roll off angle. The roll off angle is the angle at which a surface, like the leaf, must be tilted in order for the drop to roll off. With lower adhesive force the drop will roll of the leaf at a smaller "tilt angle." Other plants that exhibit superhydrophobic tendencies will have a higher tilt off angle because the area of contact is larger.

Figure 4.5 depicts the stages of interactions between the water and leaf surface. There will always be a layer of air trapped in between the two surfaces. As water comes in contact with the leaf, different forces can come into play. If the water droplet is only lightly resting on the leaf, such as a mist or dew drop, then the drop will only contact the tallest papillae resulting in minimal areal contact. When the water impinges on the leaf with greater force, for example a rain drop, then the situation shown in Figure 4.5 (a) will result. The water will be resting on the papillae with a layer of air trapped between the water and the leaf surface. Acting surface tension in the droplet will create a repulsive force on the papillae with a greater contact area, shown as "re" in the figure. When the water droplet recedes, that is, draws back from the surface, the contact areas are reduced, and the contact is released from the papillae one by one resulting in Figure 4.5 (b) where "ad" represented the existing adhesive force. Finally, in Figure 4.5 (c) the water continues to recede from the surface and is only in contact with the tallest papillae. The fact that the papillae are of different heights and therefore have a slightly different force interaction with the water means that as the droplet leaves the leaf the adhesive force is very small. Note that the cohesive force between the water molecules is strong and the water are drawn to other water molecules rather than the surface. Figure 4.5 (d) is an example of a man made hydrophobic surface.

The applications for superhydrophobic surfaces or coatings is substantial and includes self-cleaning windows and other surfaces, ship and marine coatings as well as agricultural applica-

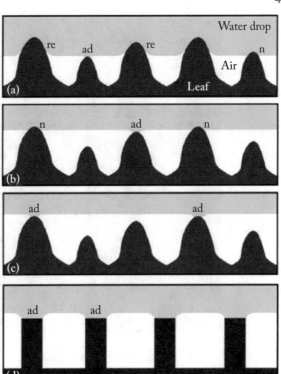

Figure 4.5: Drawing of a water droplet on the leaf surface under different conditions.

tions. To address the market desire for superhydrophobic surfaces various manufacturing approaches are being investigated. These include semiconductor fabrication methodologies, 3D printing, and biotechnology-based manufacturing processes. These processes, using various materials, often replicate the papillae covered leaf surface very well. However, the manufactured papillae are often of the same height; see Figure 4.5 (d). Therefore, the adhesive forces are uniformly in place over the surface without the gradual reduction of adhesive force found in the natural lotus leaf.

One final distinction makes the lotus leaf unique among the superhydrophobicity exhibiting plants. That is in the novelty of the wax crystals that cover the tubules on the papillae.

Figure 4.6 shows SEM images of the wax crystals from the lotus leaf upper surface (a) and the under surface (b). The remaining images are from plants with similar water interactions, spurge, yucca, cabbage, and eucalyptus, respectively. An important aspect of the figures is that each section represents the same surface area of $4 \times 3 \ \mu m^3$. Comparing the different sections in the image, the differences in density and shape of the wax crystals can be observed. In addition to the different densities and shapes, research has shown that the wax crystals also have differ-

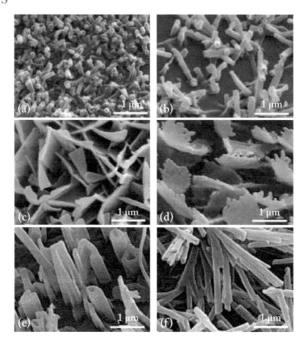

Figure 4.6: Image of the wax crystals from several plants.

ent chemical compositions. These variations result in different water capillary properties, and as such, serve as additional distinguishing characteristics among the various plants exhibiting superhydrophobic properties.

Here, superhydrophobic response of the surface of a lotus leaf was discussed. Often, as a water droplet rolls off the leaf it will collect dust, dirt, and debris from the surface. The structure of the leaf has been studied and replicated with potential applications for self-cleaning surfaces such as windows. This interaction is due to the fact that the force of attraction between a dirt particle and the water drop is stronger than the adhesive force of the particles to the lotus leaf. In this manner the dirt adheres to the water and is removed as the water rolls off the leaf.

In addition to some of the manufacturing issues that are associated with the replications of the leaf structure it is also noted that the interaction requires that an air layer is trapped between the leaf and the water. In addition, water is clearly the fluid of choice in this interaction.

Because various application may use fluids other than water, surfaces with oleophobicity are desired. That is, surfaces that tend to repel oil, instead of water are desired. Also, because a fluid other than water may have different physical properties such as viscosity, density, and surface tension, there may not exist an air layer between the fluid and the surface as was found with the lotus leaf.

Figure 4.7: The pitcher portion of the pitcher plant, species *Nepenthes muluensis*. The pitcher is the result of the modification of the end of a leaf of the plant.

Research continues, not only in the investigation to further understand the nanoscale structures of these plants resulting in unique attributes, but also regarding replication of what nature has already perfected. The results of these research efforts will contribute to new materials and products.

4.2 THE PITCHER PLANT

The lotus plant and its leaves are not the only flora that has raised curiosity and intrigued scientists for many years. The pitcher plant, with the "pitcher" shown in Figure 4.7, is an example of a plant found in abundance in many parts of the world, from Asia where the species in the figure is found, to multiple locations in North America.

A question that has been pondered by naturalists since the 1800s is "how or why do insects fall into the pitcher?" and "why do they not climb out?" Finding and understanding the answer to these questions has led researchers and engineers to develop new liquid-repellent coatings for fuel transport and biomedical applications as well as anti-fouling and anti-icing technologies.

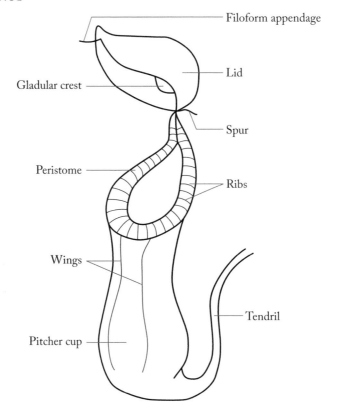

Figure 4.8: A drawing showing the various parts of the pitcher.

As shown in Figure 4.8, the part of interest with regard to capturing the insects is the peristome, the rim of the pitcher. The desired properties or effects of this rim is that an insect landing on or stepping on the rim would end up captured in the lower portion of the pitcher.

The pitchers are a modification at the end of the tendril of the leaf of the plant. They start out as small nodules and as time progresses begin to take the shape of the pitcher. The newly forming pitcher shown in Figure 4.9 is about a month "old" and less than an inch long.

The plant accomplishes trapping of insects not with a sticky substance on the rim, which could potentially trap the insect at the rim, but with a water surface. Many insects have a fluid coating on their feet to prevent them from sticking to surfaces. However, that coating, when the insect lands on water can result in the water being very slippery. Hence, the pitcher plant has macro and micro channels on the peristome (rim) which run perpendicular to the circumference of the rim, as shown in the drawing in Figure 4.8. The channels capture water which may land on the rim from rain, dew, or condensation.

There are several unique aspects of these channels, as shown in Figure 4.10.

Figure 4.9: **Photo of an immature pitcher. (Photo courtesy of Nathan Burmester.)**

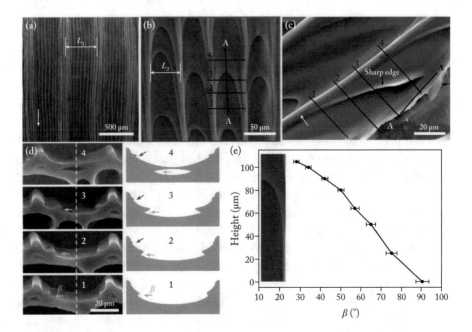

Figure 4.10: **SEM** images of the macro and micro channels in the peristome, cross section, and angle of the openings.

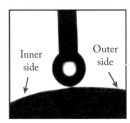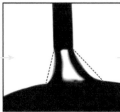

Figure 4.11: Experiment: water droplet (in white) being dropped on the peristome surface.

First, as stated above, the larger channels shown in Figure 4.10 (a) have smaller channels in them (b). Hence, a two-order structure exists with micro channels comprising the macro channels. In addition, the micro channels have several structural features. The arches or "duck-bill" features of the micro channels vary in width and angle of attachment to the edge of the micro channel (e). Finally, as represented in Figure 4.10 (d), the cross sections of the micro channels have distinct structures in them that vary with the width of the channel. This combination of factors: macro- and micro-sized channels, the arched or duck bill structure, variation in cross-section structure of the arches, and the angled sides results in unique water-controlling aspects.

A property of the peristome material is that it is superhydrophilic, that is water loving, as shown in Figure 4.11. There is essentially no contact angle when the drop of water (in white), is placed on the peristome surface. Therefore, water will be captured very quickly in the channels upon arrival at the plant surface.

All these unique features in the channels of the peristome of the pitcher plant result in two aspects that allow the plant to capture prey and maintain water equilibrium. The water droplets are captured in the channels and confined laterally. That is, water will remain in a channel and not migrate to adjacent channels or around the rim. Second, because of the duck-billed design of the micro channels and the smaller channel features (Figure 4.10 (d)), the water will flow only in one direction and that is toward the outer rim of the pitcher. This combination of features then means that an insect arriving on the peristome will land on a slippery surface and potentially hydroplane downward into the pitcher and that any water collected will be funneled to the outside of the pitcher and not into the pitcher potentially filling it with unwanted water.

From an application standpoint, the particular structure in the peristome which allows water to be trapped and flow in only one direction can be used to improve microfluidic devices. These devices, sometimes called "lab on a chip" depend on capillary action and other fluid control mechanisms to achieve their purpose. The fact that the water, once contained in the channel structure, creates a surface that is "slippery" implies that this approach may be used for anti-fouling applications where it is desired to prevent the formation of biofilms or bacteria adhesion. The structure also has application to non-water fluids since the resulting surface is oleophobic and does not require the layer of air to achieve desired responses. In this way a

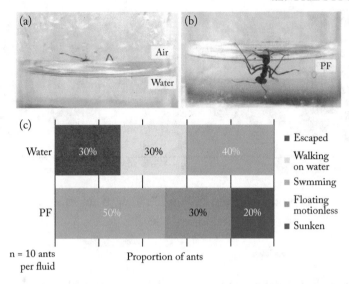

Figure 4.12: Photo and results of a study of ants trapped in the internal portion of the pitcher plant.

deposited lubricating fluid could be kept in place, resulting in micro scale slippery junctions and surfaces as desired.

Once the insect has fallen into the pitcher, the next question is "How are they kept from escaping?" The initial thought is that there must be a sticky material on the ends of the hairs or stomata on the inside of the pitcher structure. In reality, the plant secretes a viscoelastic fluid that traps the insects more effectively than water and keeps them from escaping.

An ant trapped in the pitcher fluid (PF) and the results of observation of ants interacting with water and the PF are presented in Figure 4.12.

Understanding the exact composition and physical properties, such as the viscosity, of the pitcher fluid has helped to answer the question of how the insect remains trapped within the plant. The pitcher fluid has a lower surface tension than water, allowing the insect to sink into the fluid. It was found that more energy is required to separate the insect from the fluid than on water. These properties have potential in medical diagnostic tests where it is desired to capture specific entities. In addition, the fluid strongly resists de-wetting, leaving behind a residue on the insect that encourages re-wetting. In this manner an entity that has come in contact with such a fluid, even if removed from the fluid, will more quickly become engulfed in the liquid once reintroduced. Again, this property may find applications in micro-fluids, medical diagnostics and even some manufacturing processes.

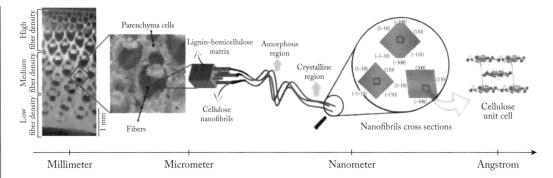

Figure 4.13: Bamboo structure from millimeter to angstrom.

4.3   BAMBOO

One of the fastest-growing plants in the world is bamboo, and for many years it has been investigated for use as a truly renewable resource. Currently, bamboo has found use in clothing, cutting boards, flooring, and reusable cleaning clothes and traditional paper products. Use of bamboo in applications traditionally filled by animal or wood products can contribute to the reduction of the carbon footprints of many products. However, the current applications for bamboo require low values of mechanical properties such as strength, response to strain, and stress. Bamboo could provide a viable material for construction, automotive, and other vehicle applications if some of the mechanical properties could be strengthened to complement other favorable attributes.

Researchers at the University of Maryland have recently developed a two-step manufacturing process that significantly increases the strength of bamboo. The process consists of a step which modifies the alignment of some of the fibers followed by microwave heating.

A lignin is a member of a class of organic polymers. These polymers are an important contributor to the structural integrity of many plants and are an important component of cell walls. They can be found in the bark and wood where they provide rigidity.

Figure 4.13 shows the various physical structures of the bamboo plant at different size scales. Throughout the years various approaches have been studied to increase the applicability of this renewable resource particularly in the area of the physical properties.

Some of the most recent research has focused on the micrometer to nanometer scale as shown in Figure 4.14. The preferred process involves two steps. The first step involves a chemical process of the bamboo called delignification. Delignification is the separation, or sometimes disintegration, of the combined lignin and cellulose structure into the fibrous components. When delignification occurs in homes built from wood it can have a devastating effect. In the process shown in the figure only a partial delignification is desired to partially remove the lignin and hemicelluloses from the cell walls causing them to become porous and softer. Because of the increased porosity, the next step—microwave heating—is used to successfully drive out mois-

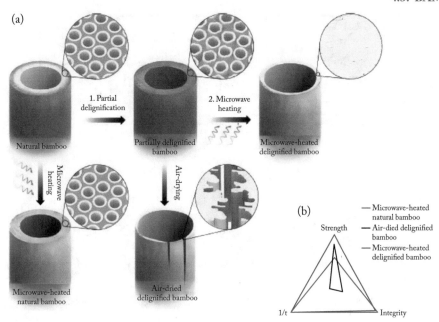

Figure 4.14: The bamboo strengthening process and variations (a) including a comparison of drying time, strength, and integrity (b).

ture. Removing the moisture causes the remaining material to shrink substantially and become intertwined resulting in densely packed cell walls. Figure 4.14 represents the bamboo structure with different variations of the process steps as well as a comparison of strength, integrity, and the inverse of drying time of the resulting bamboo structure in section (b).

In addition to having a goal for a renewable source of materials, other desirable attributes include low cost and meeting the requirements of a desired application. Contributors to the cost of a material include multiple factors such as the initial harvesting cost, transportation, and manufacturing. The cost of manufacturing is dependent upon machinery required, operating requirements such as space and utility expense as well as the complexity of the manufacturing process. Complexity is indirectly dependent on the number of steps in the manufacturing process. It is necessary also to combine the two mentioned aspects: application physical requirements and manufacturing cost. The results of this comparison are shown in Figures 4.15 and 4.16.

Sections (a) and (b) of Figure 4.15 compare tensile strength vs. strain, tensile strength and modulus for the different processing approaches. These results show that the two-step process of delignification and microwave heating has the best property values when compared to the other process approaches studied. Sections (c) and (d) represent a comparison of various physical attributes when two-step processed bamboo is compared with other materials used in similar applications.

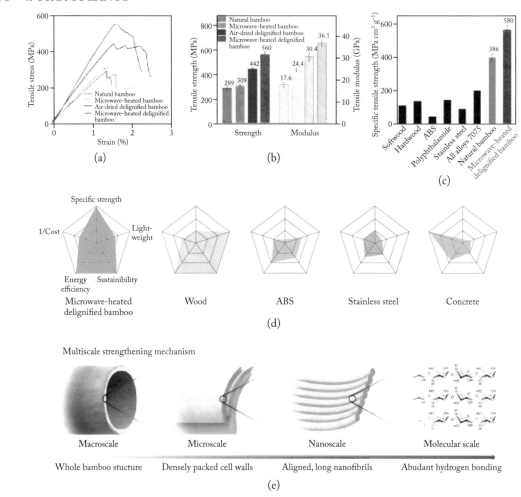

Figure 4.15: Graphical representation of the properties of the strengthened bamboo and size scale.

The desired values for all the parameters are expressed as the furthest out points on the radar chart. Hence, the microwave-heated delignified bamboo matches wood in energy efficiency and sustainability but exceeds wood in specific strength. The charts show that bamboo has higher levels of the desired attributes of stainless steel in all five areas. Comparison charts of this type are critical in assessing and defining various attributes or penalties for any new product or approach.

For multiple applications, the flexural strength, which is similar to the tensile strength of the material, is also of importance. As the name implies, the flexural strength of a material is a

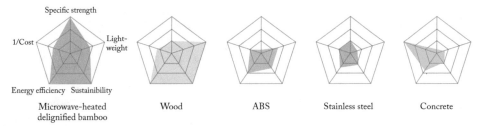

Figure 4.16: Enlargement of the radar charts comparing microwave-heated delignified bamboo.

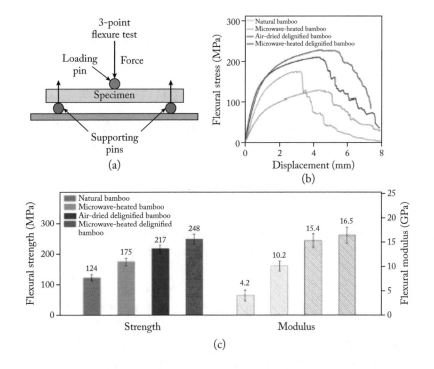

Figure 4.17: Flexure performance of bamboo following different treatment paths.

measure of how much curvature, i.e., bending when a point force is applied, can be withstood before the material under test breaks. Importantly, this test can represent the number of defects present in a material, since the material piece under test will fracture at a defect location. A material that is homogeneous with few defects will show a higher fractural strength value than one with defects present.

This figure shows the typical experimental set up used to test for flexural strength in section (a), a graph of the stress applied vs. displacement for the various bamboo treatment options (b), and in section (c) the strength and modulus. The modulus is a measure of the rise over the run

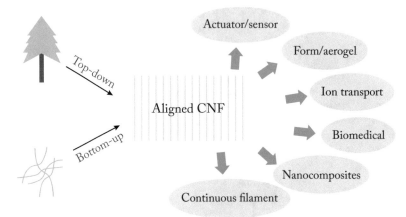

Figure 4.18: Multiple applications of cellulose nanofibers.

calculated in the linear part of the stress vs. displacement or stress vs. strain curve for Young's Modulus. The larger the modulus value the smaller the amount of displacement for a given stress applied.

The continued investigation into the use of plant materials such as lignin and cellulose for construction and vehicle use shows promise in the search for sustainable and environmentally friendly solutions.

The applications for cellulose nano fibers represented in Figure 4.18 will require smaller amounts of the plant-based materials and satisfy the physical strength parameters for the applications.

BONUS BOX

Application of plant xylem for water filtration
Obtaining clean water is one of the largest problems facing much of the world's population today. Chemicals from production facilities and agriculture have increased water pollution and a present a continuing challenge for water treatment facilities. In this case point of use filtration provides an additional level of purification.

For many without access to water treatment, viruses, bacteria, and microbes due to animal and human contamination result in water that is dangerous to use or consume. Multiple organizations are working to provide clean water to these affected regions but often the solutions are expensive or difficult to provide to the point of use.

Because many of the structures found in plants are at the micro and nanoscale research is investigating the potential to use naturally existing materials as a water filtration method.

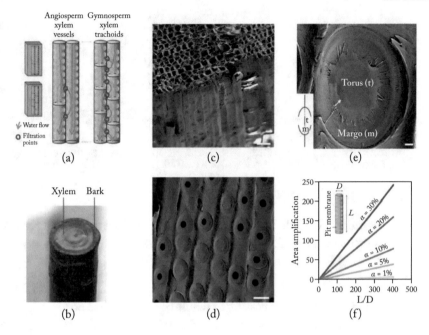

Figure 4.19: Overall structure of xylem.

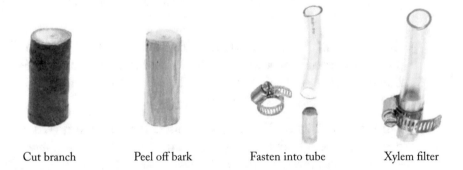

Figure 4.20: Process for constructing a xylem based filter.

Figure 4.19 shows the structure of xylem in flowering plants and other flora which may be used to filter water. Xylem is the tissue in plants responsible for transporting water. Section (a) shows different potential paths either inherent in the plant or with portions of the plant stacked. In the stacked approach on the right side of the drawing, water will go through multiple filtration points. Sections (c), (d), and (e) are various images of cross sections of the tissue. The scale bars are 40, 20, and 1 μm, respectively.

The steps to attach a piece of the plant stem to a tube which could be attached to a pump to force water through the plant-based filter and shown in Figure 4.20.

This approach may provide a path for the use of multiple types of plants with a xylem core for water filtration with applications beyond existing contaminated water systems. It may benefit communities that experience temporarily contaminated water due to natural disasters. There is potential for use widespread as a point-of-use filtration device.

4.4   REFERENCES

[1] Boutilier, M. S. H., Lee, J., Chambers, V., Venkatesh, V., and Karnik, R. Water filtration using plant xylem. *PLOS ONE*, 9(2):e89934, 2014. https://doi.org/10.1371/journal.pone.0089934 DOI: 10.1371/journal.pone.0089934.

[2] Chen, C., Li, Z., Mi, R., Dai, J., Xie, H., Pei, Y., Li, J., Qiao, H., Tang, H., Yang, B., and Hu, L. Rapid processing of whole bamboo with exposed, aligned nanofibrils toward a high-performance structural material. *ACS Nano*, 14(5):5194–5202, May 26, 2020. DOI: 10.1021/acsnano.9b08747.

[3] Chen, H., Zhang, P., Zhang, L. et al. Continuous directional water transport on the peristome surface of nepenthes alata. *Nature*, 532:85–89, 2016. https://doi.org/10.1038/nature17189 DOI: 10.1038/nature17189.

[4] Ensikat, H. J., Boese, M., Mader, W., Barthlott, W., and Koch, K. Crystallinity of plant epicuticular waxes: Electron and X-ray diffraction studies. *Chemistry and Physics of Lipids*, 144(1):45–59, 2006. https://doi.org/10.1016/j.chemphyslip.2006.06.016. DOI: 10.1016/j.chemphyslip.2006.06.016.

[5] Ensikat, H. J., Ditsche-Kuru, P., Neinhuis, C., and Barthlott, W. Superhydrophobicity in perfection: The outstanding properties of the lotus leaf. *Beilstein Journal of Nanotechnology*, 2:152–161, 2011. https://doi.org/10.3762/bjnano.2.19 DOI: 10.3762/bjnano.2.19. 85

[6] File:Nepenthes in the Southern Western Ghats.jpg. (2021, September 28). Wikimedia Commons, the free media repository. Retrieved 04:12, October 15, 2021. https://commons.wikimedia.org/w/index.php?title=File:Nepenthes_in_the_Southern_Western_Ghats.jpg&oldid=594156096

[7] File:Nepenthes pitcher morphology upper.svg. (2021, September 4). Wikimedia Commons, the free media repository. Retrieved 04:09, October 15, 2021. https://commons.wikimedia.org/w/index.php?title=File:Nepenthes_pitcher_morphology_upper.svg&oldid=588586783

[8] Kang, V., Isermann, H., Sharma, S., Wilson, D. I., and Federle, W. How a sticky fluid facilitates prey retention in a carnivorous pitcher plant (nepenthes rafflesiana). *Acta Biomaterialia*, 128:357–369, July 1, 2021. DOI: 10.1016/j.actbio.2021.04.002.

[9] Koch, K., Dommisse, A., and Barthlott, W. Chemistry and crystal growth of plant wax tubules of lotus (Nelumbo nucifera) and nasturtium (Tropaeolum majus) leaves on technical substrates. *Crystal Growth and Design*, 6(11):2571–2578, 2006. DOI: 10.1021/cg060035w.

[10] Li, J., Zheng, H., Yang, Z. et al. Breakdown in the directional transport of droplets on the peristome of pitcher plants. *Communications Physics*, 1:35, 2018. https://doi.org/10.1038/s42005--018-0038-z DOI: 10.1038/s42005-018-0038-z.

[11] Li, K., Clarkson, C., Wang, L., Liu, Y., Lamm, M., Zhenqian, P., Zhou, Y., Qian, J., Tajvidi, M., Gardner, D., Tekinalp, H., Hu, L., Li, T., Ragauskas, A., Youngblood, J., and Ozcan, S. Alignment of cellulose nanofibers: Harnessing nanoscale properties to macroscale benefits. *ACS Nano*, 15, 2021. DOI: 10.1021/acsnano.0c07613.

[12] Wong, T. S., Kang, S., Tang, S. et al. Bioinspired self-repairing slippery surfaces with pressure-stable omniphobicity. *Nature*, 477:443–447, 2011. https://doi.org/10.1038/nature10447 DOI: 10.1038/nature10447.

[13] Youssefian, S. and Rahbar, N. Molecular origin of strength and stiffness in bamboo fibrils. *Scientific Reports*, 5:11116, 2015. https://doi.org/10.1038/srep11116 DOI: 10.1038/srep11116.

CHAPTER 5

From Birds

The avian world offers unique aspects of feather structures resulting in colors that stand alone in the natural world. Some birds also have a different structural approach that allow for beaks of amazing strength. Each of these observed attributes is due to nanoscale structures. Applications may provide positive impacts at a global level.

5.1   TOUCAN BEAKS

Not only does the Amazon shelter many unique species of plants and reptiles, but it also is home to the bird family Ramphastidae of the Class Aves which includes over 40 different species. Probably the most familiar is the toco toucan shown in Figure 5.1.

It is obvious from the photo that the beak of the toucan comprises about 1/3 of its body. However, the beak only comprises approximately 1/20 of the mass of the bird. Because of this mass distribution the toucan is allowed to soar through the tops of the forest trees with minimal effort. From the observations of explorers, it was also noted that the beak of the toucan was strong, having observed the birds using them as weapons. There was a disconnect between the two observations: the beaks did not seem to be flexible or flimsy, yet the bird had no difficulty flying. The majority of the time, if something is observed to be strong or unbendable, it is assumed that it is solid and heavy. In most birds, beaks are only a very small portion of the total size (length) of the bird. For example, the American robin averages a length of 10 inches, yet the beak is merely 0.5-inches long. So, a denser beak for a robin would not hinder its flying ability. This would not be the case for a toucan, where a heavy beak that is 1/3 of its length would impact flight.

Therefore, the observation of a toucan in the Amazon Forest raised the question of the composition of the beak that provided strength and the ability to fly. At a macroscopic level, the beak appeared to be similar to the beaks of other birds and made up of keratin, a protein, arranged in layers. The layers of keratin, over the bone gives bird beaks their glossy surface and the keratin is replenished as it wears down.

It was not until early this century that researchers were able to use the tools of nanotechnology to study the structure of the beak of the toco toucan. The resulting observed structure is shown in Figure 5.2. As shown, the beak contains the outer keratin layer covering a horizontal dermis layer which in turn covers a foam region comprised of bone structures and hollow areas.

Using an SEM, the detail of the "foam" structure could be observed, as shown in Figure 5.3. Most of the cells have membranes sealing them, similar to a drum structure, surrounded

Figure 5.1: The toco toucan found in the Amazon region.

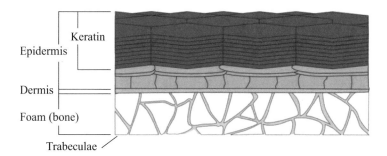

Figure 5.2: Drawing of the cross section of the beak of the toco toucan.

by the struts. Even though an SEM was used to capture these images the physical structures creating the foam are still "large" when considering the nanoscale. The scale bars in Figure 5.3 are in microns or 10^{-6} m. One thousand nanometers is equal to 1 μm. Although this foam inner structure may not officially qualify as nanotechnology there are two aspects for consideration. First, it required nano-based "scopes" to determine and define the structure. Second, and perhaps most important, is the fact that the foam structure that results in a relatively strong structure but does not have the amount of weight usually found in a structure of this size. Therefore, the potential to replicate the structure with existing manufacturing processes that operate at the micro scale is increased.

Current research into the manufacture of foam structures of this type has predominantly used a metal as the material. In the toucan's beak it is a biological material, bone, that creates the structure around the hollow regions.

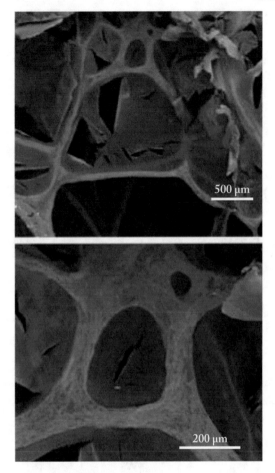

Figure 5.3: SEM images of the structure of the beak.

In comparison, metal, currently used to create a lightweight yet structurally strong material can be expensive and often the process requires a higher manufacturing temperature. The composition of the three layers of materials making up the toucan beak structure is shown in Figure 5.4.

Note that the keratin and membranes have a high concentration of carbon, which is representative of a protein-based structure. They are similar in composition with a slight variation in sulfur and potassium concentrations. The component labeled as fibers in Figure 5.4 shows a high concentration of calcium with reduced amount of carbon representative of a change in the protein composition. Continued investigation included further understanding of the specific structures within the beak and the impact of those structures on the physical properties of the beak as a whole.

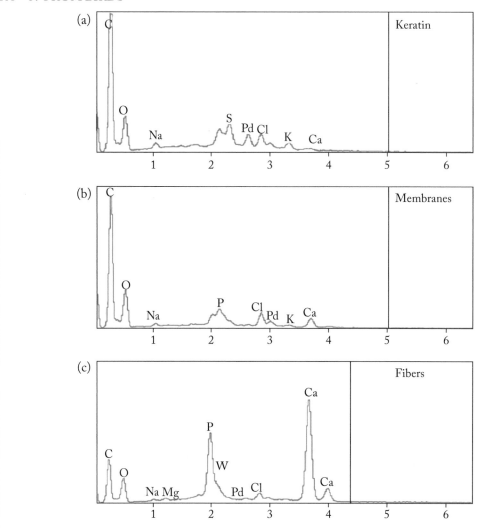

Figure 5.4: Composition of the beak structures determined by energy dispersive X-ray testing.

The outer layer of keratin as show in the drawing in Figure 5.2 is composed of horizontal, hexagonal-shaped scales that are separated by a biological-based, flexible material. Keratin represents a group of usually high sulfur compounds that form filaments 3–7 nm in diameter. These filaments are usually embedded in an amorphous matrix and are found in hair, nails, claws, fur, and many other animal structures. This structure of the toucan beak keratin is similar to what is found in the shell of the abalone. The combination of the rigid scales and the more flexible mortar or glue provides a tensile strength of about 50 MPa and a Young's Modulus of 1.4 GPa.

Tensile strength is a measure of the resistance of a material to breaking when a strain is applied or the ability of a material to "hold together" under an applied strain. Polypropylene, in the plastic family of materials, has a tensile strength of around 13 MPa, whereas stainless steel has a value of 520 MPa. Young's Modulus is also a mechanical property and is a measure of the elasticity of a material. It is represented by the symbol E and is the ratio of the stress acting on a material to the resulting stress. The Young's Modulus value for many foams is less than 0.05 GPa, that of rubber materials < 0.1 GPa and wood and wood products have an E value from 0.08 GPa up to 25 GPa. The hardest material, diamond, has a Young's Modulus value of 1210 GPa. This keratin structure has a strain-rate sensitivity because of the slippage that occurs between the scales due to the flexibility of the glue. The strain-rate sensitivity, as the name implies, is a measure of the response of a material when strain is applied to that material at different rates. For example, some materials may show an elastic region property when strain is applied slowly and a different, perhaps more brittle, response when the strain is applied more abruptly. Consider the response of Silly Putty™ to pulling, slowly or abruptly.

Although the keratin layer of the beak has been tested and measured extensively, basically because it is a continuous membrane that lends itself to multiple types of physical property measurements and determinations, it does not hold all of the explanation for the unique properties of the toco toucan beak.

It has been found, through direct testing of the entire beak and computer simulation, that it is the combination of the keratin scale/glue shell structure and the underlying foam structure that creates the unique entity that is the beak of the toucan. Seki et al. [10] has shown that the underlying foam provides an elastic foundation and helps to stabilize the deformation stresses on the beak. It is not only the addition of the shell and the foam, but the fact that the foam fills the shell is a combination that results in improved mechanical stability and the capacity to absorb energy.

The strength and adaptability of this combined structure of a layer of scales or flat pieces with a flexible glue overlaying a foam base with structurally sound struts has multiple potential applications. These applications include toys, propeller blades, vehicles, and appliance housings. Currently, some of these applications use foam like structures that are fabricated using various metals. Research that could lead to use of biological materials, such as proteins, may reduce weight and cost. Similarly, applications that currently use solid sheets of metal or plastic, or cases where metal foam is included in a metal pipe (Al is often used) could be improved by using the combination not only of materials but also structures found in the beak of the toco toucan.

5.2   THE NATURE OF FEATHERS

The next two sections deal with coloration observed in the feathers of peacocks and humming-birds. In order to better understand the results of the research and the ramifications of structural

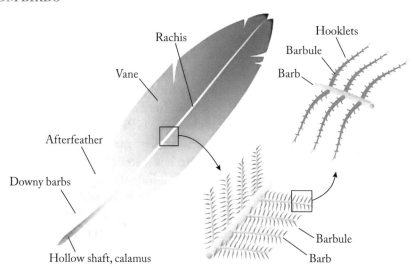

Figure 5.5: The anatomical structures in a feather.

differences this introductory section discusses some of the definitions required for this understanding.

The various portions and anatomical names are shown in Figure 5.5. Although the overall length and width of the feather will vary dependent upon the species, in many cases the
size of the barbs, barbules, and hooklets will be similar with mm and nm structures. In the
following sections, the research reviewed has shown that although many feathers have similar
structures and materials, it is the size and arrangement of those materials that results in the reflected wavelengths of light and the associated colors that are seen. Figure 5.6 shows how the
physical structure of the barbules (cross section) and the different shapes of the melanosomes for
different bird families. Melanosomes are granules produced by melanocytes which are covered
in a membrane. Melanosomes are often found in the skin and are the carriers and transporters of
melanin, a pigment. Of note here is that the melanosome shape for the peacock feather is a solid,
round, rod-type structure and the hummingbird structure is flattened, hollow, and oval-shaped.

Figure 5.7 represents the aspects of coloration both in the skin of bird (a) and in the
feathers (b) through (e). The left-hand side of the figure shows the structure of the bird skin and
the location of the melanosomes. Of interest for this section is the right side of the figure. The
keratin material can vary and is classified as an alpha ketratin and a beta keratin. Shown here
is the beta keratin which is found in birds and reptiles and occurs in more of a sheet structure.
{Alpha keratin is found in mammals and has a coiled structure. Alpha keratin is composed of
amino acids and forms hair, nails etc.} As shown in the figure, within a cell, where the boundary
is shown as the vertical dark line (d), the inner cell liquid cytoplasm contains the melanosome.

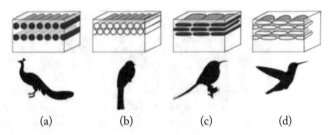

Figure 5.6: Cross section of barbules in various families showing the various melanosome structure and organization.

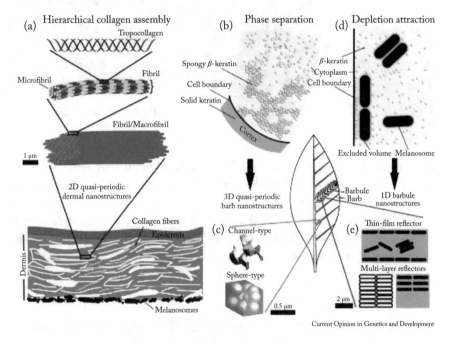

Current Opinion in Genetics and Development

Figure 5.7: Various anatomical aspects of avian skin and feathers.

The melanosome are cellular structures that provide pigment to skin and the cortex in birds and reptiles.

The right side represents the structure of the feathers. Note in the bottom-right portion of the figure that the melanosome can be solid or hollow, as shown in the previous figure and in different arrangements which will modify the interaction of the melanosomes in the barbule with light.

The barbs of white bird feathers have been studied at a scale of the micrometer [12]. This study was investigating the near infrared reflectance of white feathers and determined that the

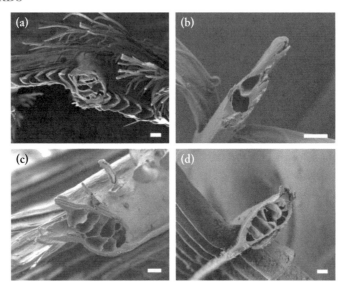

Figure 5.8: SEM images of the barbs of white feathers. The scale bars are 10 μm.

average and relative reflectance could be correlated to the barb structure of the feather. Figure 5.8 shows SEM images of four feather barbs. Sections (a) and (c) represent lower values of the two quantities than (b) and (d).

The research has showed that even at the microscale, multiple aspects of the structure of feathers will result in different reflectance values and are dependent on the wavelength of the interacting light. Also, the structures differ among species of birds in the same family.

5.2.1 PEACOCK FEATHERS

Peacocks are certainly acknowledged as one of the most colorful in the animal kingdom. The multitude of colors that are observed and the generation of those colors has fascinated researchers for hundreds of years. Again, it was not until the last few decades that we have had the tools to investigate the nanoscale structure of the feathers that has allowed the understanding of the various color creation.

Figure 5.9 is a photograph of one of the "eyes" in the tail of a male peacock. The transition of colors from purple to blue to brown to green along what appears to be the same vane has remained a mystery until recently.

The colors that are observed are a result of visible light wavelengths interacting with physical structures and not chemical pigments, as was discussed in Section 1.1, and can be due to several types of interactions. These interactions include thin film interference and coherent or incoherent scattering. In turn, the type of interaction, the wavelength (color) of light impacted

Figure 5.9: A photograph of the "eye" on a male peacock feather.

and the observed effects are dependent upon the material composition, shape, and thickness. As seen previously, it is also dependent upon the number of intervening layers of material.

This description above represents the multitude of interactions and dependencies that result in the rainbow of colors observed in various bird feathers. It is important to note that the degree of interaction (structural coloration) of a given color with a particular structure is dependent upon the difference between the size of the structure and the wavelength of light. In Section 3.3 it was discussed that when the structure size was smaller than the wavelengths of visible light that the structure became anti-reflective. For the peacock feather, since colors are observed, it requires that the material have the same order of size as the wavelength. Hence, since visible light has wavelengths of the range of 400–700 nm, it is implied that the physical structures impacting the absorbed and reflected light is of a similar size.

Recall from Figure 5.7 that the barbules on feathers can have different structures. The result of the different structures is that different sections of the feather will reflect different wavelengths of light, hence, different colors. The exact mechanism that creates the multitude of colors along a single peacock feather is a result of several interactions including thin film interference and scattering. It is necessary to consider the nanoscale structure to understand the mechanisms.

The set of SEM images in Figure 5.10 shows different regions of the feather. Section (a) in the figure is a cross section of the outer cortex of the barbule in the feather portion that reflects

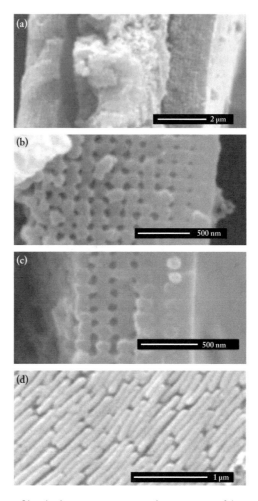

Figure 5.10: SEM images of barbules structures in the green and brown regions of the peacock feather.

green. Sections (b) and (c) are a higher magnification of the barbules that reflect green and brown, respectively. The surface of the cortex is a keratin sheath and beneath that layer is a region of melanin rods connected by keratin that comprise a 2D photonic crystal-like structure. Note that sections (b) and (c) represent a view of the melanin rods from a cross sectional, perspective. The dark gray regions in the SEM image are the hollow areas in between the rods. Although the images in (b) and (c) appear to be similar close evaluation shows that the shape and the size of the rod differs between the two images. As a result of this difference, feather portions with the different shaped barbules will reflect colors of light uniquely. Section (d) of the figure shows the melanin rods, with the keratin removed.

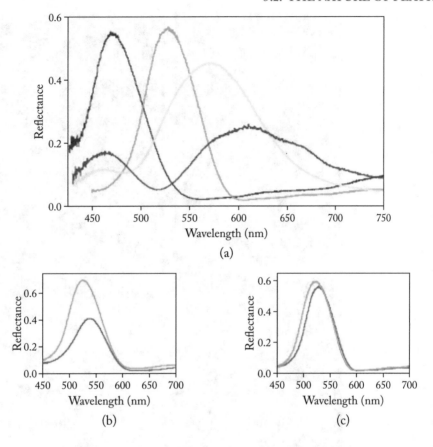

Figure 5.11: Various reflectance vs. wavelength curves for different barbule sections.

The primary differences between the regions of the feather that reflect the different colors are in the rod spaceing (lattice constant) and the number of periods or melanin rod layers lying normal to the cortex surface.

When different parts of the feather were tested for reflectance, the different portions showed different values as shown in Figure 5.11. The line colors in the upper graph represent the portions of the barbule that reflect those colors. The difference in reflectance based on wavelength clearly shows the dependence on the observed (SEM) structural differences. That is, the observed differences in the rod spacing and the number of rod melanin layers (periods) are the contributing factors to the difference in reflected colors. The lower two graphs in the figure are measurements for different polarizations. Varying the rod spacing shifts the partial photonic bandgap, restricting the wavelengths that can be reflected. Changing the period impacts the production of additional colors. This results in additive and mixed coloration.

Figure 5.12: Tail feathers of the male peacock and barbules.

Normal view and magnified images of the peacock feather are shown in Figure 5.12. The upper portion of the figure shows the feathers as they transition between colors with images in (c) and (d) showing one region with the barbules branching out of the barb and overlapping barbules (d). The lower portion of the figure are micrographs of single barbules along the colored regions of the barbule. The color bars at the bottom of portions (e) through (j) are the color in that region. Note the substantial tilt to the segments in portion (e) which is found in the violet region of the feather.

Figure 5.13 is a series of TEM images of a barbule, with the top image showing the cross section of the barbule. The lower series of images are cross sections of the six colored regions as shown by the bars along the bottom of the figure. This figure clearly shows the difference in density, diameter, spacing, and arrangement of the melanin rodlets (dark dots) and air regions

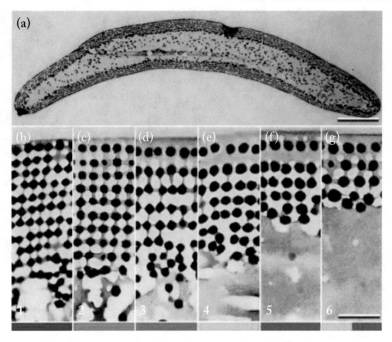

Figure 5.13: TEM images of a barbule and melanin rods in different color regions.

(light gray). The darker gray in some of the images is keratin. The scale bar in the upper TEM image is 5 μm and the lower-half scale bar is 0.5 μm (500 nm).

As with some other creatures in nature the beautiful colors in a peacock's tail are due to nanoscale structures and specific arrangement of cell organelles such as the melanosomes. By studying how the physical structures in the barbules change along the length of the peacock feather and as a result reflect different colors, new types of chemical sensors could be created. In these very small sensors color changes would occur due to structural changes, hence, detecting the presence of desired or harmful chemical elements.

5.2.2   HUMMINGBIRDS

Nature never runs out of ways to amaze, startle, and intrigue us. Hummingbirds are one of nature's most amazing creatures. Hovering in mid-air to sip nectar from flowers, heartbeats over 1,000 beats per minute and migrations of 500 miles in a single segment, these tiny avains dazzle with their colorful feathers.

As shown in Figure 5.6, the type and arrangement of the melanosomes in the barbules of birds is different among bird families. The hummingbird has melanosome structures that are flat and more oval than the circular rod-shaped structures found in the peacock feathers.

Figure 5.14: The Rivoli's hummingbird.

The overview of the morphology of the hummingbird feather is shown in (b) of Figure 5.15. The center and right portions of the figure show a cut away drawing of the barb (center) and an expanded version of the barbule, showing the "boomerang" shape b.3. It is within the boomerang shape that the melanosomes of the barbules are found as in the inset above b.3 in the figure. In (c) of the figure a drawing of the flat, oval-shaped melanosome is given.

In the 1970s before extensive investigation into the structure of the feathers of hummingbirds, it was believed that most of the melanosomes were arranged as shown on the left side Figure 5.16, that is, flat, air-filled melanosomes in the keratin. More recently, researchers have found structures with melanin-filled melanosomes (middle of the figure) and combination structures with both air-filled and solid melanosomes.

Differentiation in the locations of the different melanosome structures is exemplified in Figure 5.17 with TEM images that show on the back of the hummingbird the structure is the solid melanin-filled melanosomes and in the throat region the hollow platelets are found.

The hierarchical structure of the barbules of the hummingbird are shown in Figure 5.18 with increasing magnification. Section (a) presents the barbules (*) on both sides of the barb with the lamina, sidewalls (arrowheads), and long hooks (arrow) shown. Increasing magnification of the SEM images results in (d) which shows the layer and air holes in the lamina. Finally, TEM images in (e) and (f) show again the now familiar layers of melanosomes.

These layer's structures have been simulated to study the thin film interference which can result from the multiple layers of defined thicknesses and composed of keratin, melanin, melanosome membrane, and air. Results of the simulation are shown in Figure 5.19.

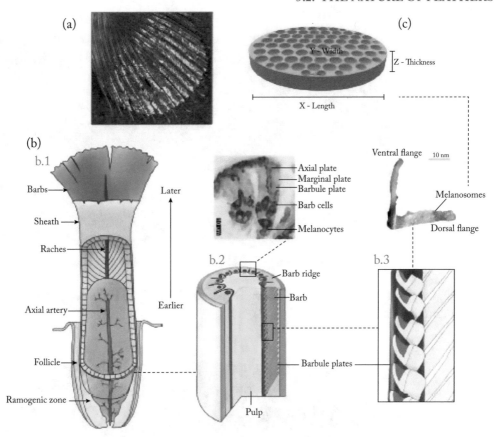

Figure 5.15: The physical structure of the barb on a hummingbird feather with additional levels of detail provided.

Graphs in (a) and (b) represent the reflectance dependence with varying thickness in the top melanosome layer, b.1 and the thickness of the air layer, b.2. In section (a), b.1 was varied from a thickness of 20–50 nm, and b.2 was kept at a thickness of 110 nm. In graph (b), b.1 was held constant at 25 nm while b.2 was varied from 110–125 nm. Section (c) shows the dependence of reflectance on the number of periods and (d) varies the angle of incidence. Each of these variations show some wavelength dependence on reflectance. The mean values, shown as the thicker red line in the graphs, are a good match with the experimentally observed values for the gorget of Anna's hummingbird.

Fascination with the avian world has led to many discoveries of the similarities among the group such as the mechanism of very light feathers and bone structures. At the same time, the variety of nanoscale structures that contribute to unusual strength as in the toucan beak and an extensive rainbow of colors in peacocks and hummingbirds leaves room for further investigation.

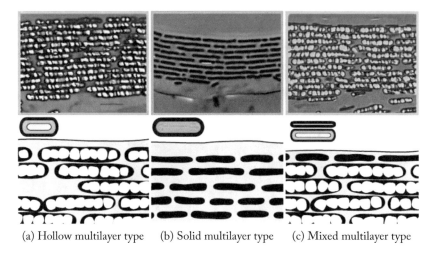

(a) Hollow multilayer type   (b) Solid multilayer type   (c) Mixed multilayer type

Figure 5.16: SEM images and drawn representations of the different melanosome structures in hummingbird barbules.

Figure 5.17: Variations in the melanosome structures are shown for the white-booted racket-tail hummingbird.

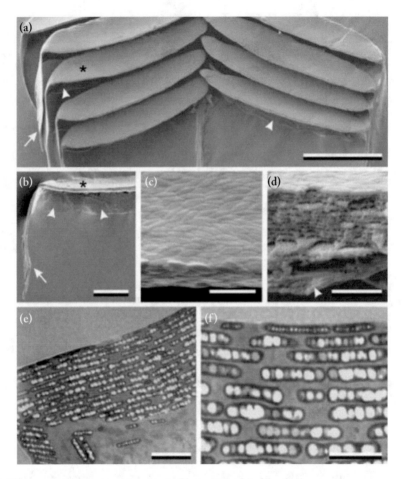

Figure 5.18: Electron microscopy images of hummingbird barbules.

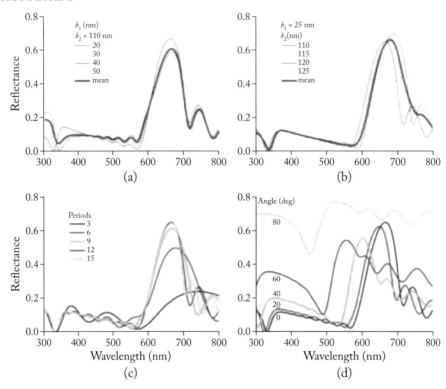

Figure 5.19: Resulting reflectance spectra based on varying parameters within the barbule structure.

These discoveries have the potential to result in improved photonic devices, enhanced cosmetics and paint materials, decorative and functional coatings, and improved communication and sensing systems.

5.3   REFERENCES

[1] Beltran Segura, D. F. Color reflectance in two iridescent Andean hummingbirds and its nanostructural basis, January 2016. https://repositorio.uniandes.edu.co/bitstream/handle/1992/18738/u722324.pdf?sequence=1

[2] D'Alba, L., Meadows, M., Maia, R., Yeo, J. S., Manceau, M., and Shawkey, M. D. Morphogenesis of iridescent feathers in Anna's hummingbird calypte anna. *Integrative and Comparative Biology*, icab123, 2021. https://doi.org/10.1093/icb/icab123 DOI: 10.1093/icb/icab123.

[3] Freyer, P. and Stavenga, D. G. Biophotonics of diversely coloured peacock tail feathers. *Faraday Discuss*, 223:49–62, 2020. http://dx.doi.org/10.1039/D0FD00033G DOI: 10.1039/d0fd00033g.

[4] Giraldo, M. A., Parra, J. L., and Stavenga, D. G. Iridescent colouration of male Anna's hummingbird (Calypte anna) caused by multilayered barbules. *Journal of Comparative Physiology A, Neuroethology, Sensory, Neural, and Behavioral Physiology*, 204(12):965–975, 2018. https://doi.org/10.1007/s00359--018-1295-8 DOI: 10.1007/s00359-018-1295-8.

[5] Gruson, H. Origin and functions of iridescent colours in hummingbirds, 2019. https://www.researchgate.net/publication/339067556_Origin_and_functions_of_iridescent_colours_in_hummingbirds

[6] Kazilek, C. J. Feather biology. *ASU—Ask a Biologist*, August 11, 2009. https://askabiologist.asu.edu/explore/feather-biology

[7] Meyers, M. A., McKittrick, J., and Chen, P. Y. Structural biological materials: Critical mechanics-materials connections. *Science*, 339:773, 2013. DOI: 10.1126/science.1220854.

[8] Nordén, K., Faber, J., Babarović, F., Stubbs, T., Selly, T., Schiffbauer, J., Peharec Štefanić, P., Mayr, G., Smithwick, F., and Vinther, J. Melanosome diversity and convergence in the evolution of iridescent avian feathers-implications for paleocolor reconstruction. *Evolution*, 73, 2018. doi: 10.1111/evo.13641. DOI: 10.1111/evo.13641.

[9] Saranathan, S. and Finet, C. Cellular and developmental basis of avian structural coloration. *Current Opinion in Genetics and Development*, 69:56–64, 2021. https://doi.org/10.1016/j.gde.2021.02.004 https://www.sciencedirect.com/science/article/pii/S0959437X21000241 DOI: 10.1016/j.gde.2021.02.004.

[10] Seki, Y., Schneider, M. S., and Meyers, M. A. Structure and mechanical behavior of a toucan beak. *Acta Materialia*, 53(20), 2005. https://doi.org/10.1016/j.actamat.2005.04.048 DOI: 10.1016/j.actamat.2005.04.048. 109

[11] Sharp, C. J. Photograph of the Toco toucan (Ramphastos toco), Rio Negro, the Pantanal, Brazil. *Wikipedia*, September 6, 2015. https://commons.wikimedia.org/w/index.php?curid=44158723

[12] Stuart-Fox, D., Newton, E., Mulder, R. A., D'Alba, L., Shawkey, M. D., et al. The microstructure of white feathers predicts their visible and near-infrared reflectance properties. *PLOS ONE*, 13(7):e0199129, 2018. https://doi.org/10.1371/journal.pone.0199129 DOI: 10.1371/journal.pone.0199129. 111

[13] Wang, M., Meng, F., Wu, H., and Wang, J. Photonic crystals with an eye pattern similar to peacock tail feathers. *Crystals*, 6(8), 2016. DOI: 10.3390/cryst6080099.

[14] Wikimedia Commons. Magnificent hummingbird, 2020. https://commons.wikimedia.org/w/index.php?title=File:MagnificentHummingbird.jpg&oldid=473332434

CHAPTER 6

The Amazing Human

The previous chapters have focused on numerous entities within the plant and animal kingdoms. Using the tools of nanotechnology, researchers have been able to find some of the answers to questions about why we observe what can be seen with our eyes and how what is observed is created. As shown, many of the structures that create what is observed are in the nanoscale size range. And humans are no different from the Morpho butterflies or the plant as a subject of curiosity.

Nanotechnology, and the tools created to observe our world at the molecular and atomic level, partnered with other technological advances such as Magnetic Resonance Imaging (MRI) and Nuclear Magnetic Resonance (NMR) have allowed for the observation and understanding of the human body. As discoveries continue to abound, an understanding is gained that supports the degree of awe that is inspired by the complexity and integration of all the "pieces."

This chapter, in contrast to the previous four chapters, starts at the nanoscale with the Central Dogma of Biology and covers the impacts of understanding DNA and proteins. With this knowledge of nanoscale understanding, the unique functional aspects of our nanoscale structures can be applied in multiple applications.

The Central Dogma of Biology is DNA encodes RNA; RNA encodes proteins. This process sounds pretty simple, but the level of interdependencies is only partially discovered.

6.1 DNA

Deoxyribonucleic acid (DNA) is the foundation and code structure of all living things. It is a relatively simple structure composed of 4 of the 20 amino acids, 1 sugar molecule, and 1 phosphoric acid molecule. It has a structure, now familiar to most, of a twisted ladder.

The four amino acids are: Adenine (A), Cytosine (C), Guanine (G), and Thymine (T). Their atomic structure is shown in Figure 6.1. The categories shown, Pyrimidines and Purines, are defined primarily based on the ring structure of the amino acids.

The Pyrimidines have one carbon (C) nitrogen (N) ring with two nitrogen atoms. The Purines, as shown in the figure, have two C-N rings with four nitrogen atoms. There are other differences in melting and boiling points and the method used to synthesize the structures. When referring to DNA structure the four amino acids are called bases. The four bases will pair with each other in a particular manner with A pairing with T and C pairing with G. The A-T bond is comprised of two hydrogen bonds and the C-G bond is comprised of three hydrogen bonds. Because of this restriction in chemical bonding adenine always pairs with thymine and

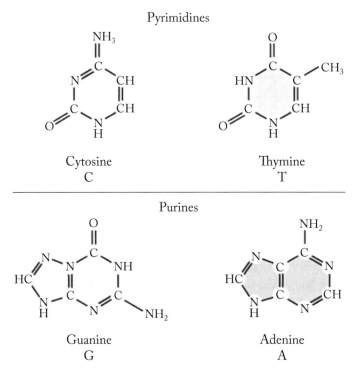

Figure 6.1: The four amino acids of DNA.

guanine always pairs with cytosine. This pairing structure results in the rungs of the ladder and are referred to as the base pairs of DNA.

The amino acids, or bases, pair up with a sugar and phosphate, as shown in Figure 6.2. This structure is called a nucleotide. The nucleotides pair up determined by the base pairing structure. The phosphate group coupled with the sugar form the backbone of the DNA strand, or in the ladder metaphor, they represent the outer supporting structure to the rungs.

The resulting structure is shown on the right side (a) of Figure 6.3 with the detailed atomic structure of two rungs shown on the left side (b) of the figure.

Of note are two aspects of the structure presented in the left side of Figure 6.3. First, is the specificity of the hydrogen bonds between the amino acids provides a level of selectivity with regard to chemical bonding. For example, if adenine is added to a solution with both guanine and thymine, the pairing would occur successfully between the adenine and the thymine creating pairs of molecules which could be separated from the solution.

Second, be aware of the negative signs near the oxygen atoms on the periphery of the structure shown in (b) of the figure. Because of the complex bonding associated with the *p* and *d* orbitals of phosphorus and oxygen atoms some of the oxygen atoms will acquire an extra

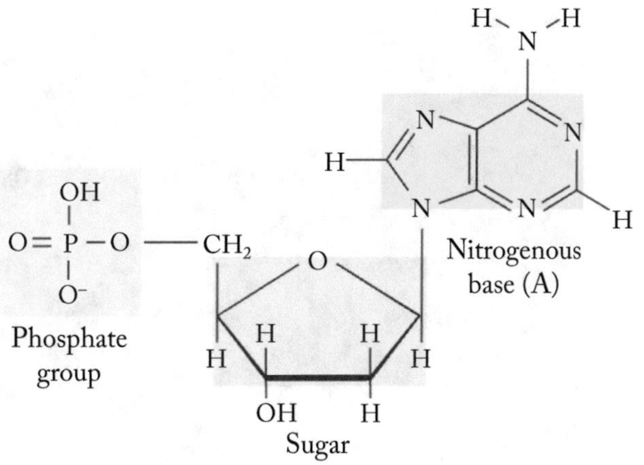

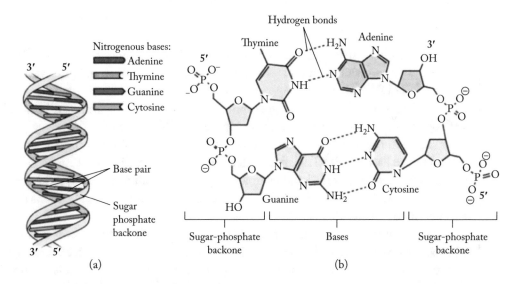

Figure 6.2: **Nucleotide with an adenine base.**

Figure 6.3: **DNA structure and nucleotide detail.**

electron and therefore carry a negative charge. These oxygen atoms are on the outside of the DNA structure; hence, DNA has an overall negative charge. It is not a charge neutral entity as can sometimes be construed based on images.

Both of these factors lead to interesting results and capabilities of DNA. Since it has an overall negative charge, it can be controlled by an electric field or used to create specific chemical interactions with other charged molecules, separate from its use in cells and the conveying of genetic information.

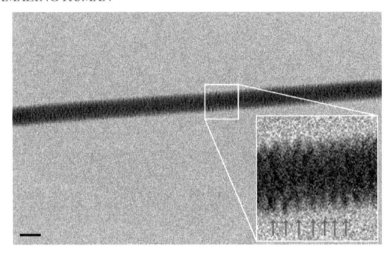

Figure 6.4: TEM image of a group of 7 DNA strands.

The laddered structure of DNA is termed a double-stranded DNA (ds DNA) and is approximately 2 nm wide with each base pair contributing between 0.34 nm and 0.68 nm to the length of the DNA, depending upon the experimental measurement method.

A TEM image of a group of seven DNA double strands arranged in a circular pattern with one strand at the center is shown in Figure 6.4. The scale bar in the image is 20 nm and the inset shows the edges of the outer strands. The red arrows point out the 2.7-nm pitch which is related to the Fast Fourier Transform calculation associated with the TEM measurement hence agreement between observed and estimated dimensions of the DNA strand bundle.

As shown in Figure 6.5, DNA is considered to have multiple applications in the diagnostic and treatment medical fields.

The multitude of potential applications are due to the attributes previously mentioned. That is, the negative charge associated with strands of DNA and that it is, indeed, a biocompatible material. Another factor that is often unmentioned is the fact that it is a simple structure as Figures 6.1 and 6.2 exemplify. When one component of the overall equation is straightforward, in this case the structure and components of DNA, the complexity of defining the mechanism and desired response for interacting with more complicated systems is simplified.

In this manner, DNA lends itself to the manipulation of complex enzymes and proteins as well as various markers. It also has been used in sensor devices when matched with photonic molecules. DNA nanotechnology as a technique used to create many structures and devices, is being viewed as a significant weapon in the fight against cancer.

Use of DNA and associated structures in the fight against cancer is a more recent endeavor, whereas a search for applications of DNA began in the semiconductor industry in the 1990s. As the need and desire for more computing power grew, the necessity to place more

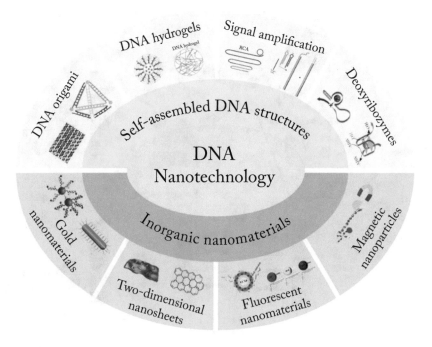

Figure 6.5: Different nanotechnology-related applications of DNA.

transistors in each unit area grew. Essentially, transistors had to become smaller. One of the changes then, and that continues today, is how to create nanometer size circuits and the wires that are equally reduced in size. The wires are needed between elements and as interconnects to the larger scale electrodes. Traditional wire manufacturing methods evolved to create wires with reduced diameters yet still too large to meet the requirements as the device size continued to be reduced.

Small numbers of atoms of metals such as silver and gold could be arranged to create a wire, but the process is time consuming and difficult. Figure 6.6 presents an alternate approach which uses a strand of DNA as a template. Because of the charge of the DNA, metal atoms can be attracted to the DNA and essentially create a conductive wire. The process is shown in the figure and begins with two gold electrodes. A DNA bridge connecting the two electrodes is created in portion (c), then the DNA attracts silver ions from the solution with the negative charge. The silver ions attach to the DNA and form a continuous wire.

DNA continues to be investigated for applications ranging from lasers and other photonic devices, to scaffold material for supporting cell and tissue growth. The building blocks of all animals continue to find new applications.

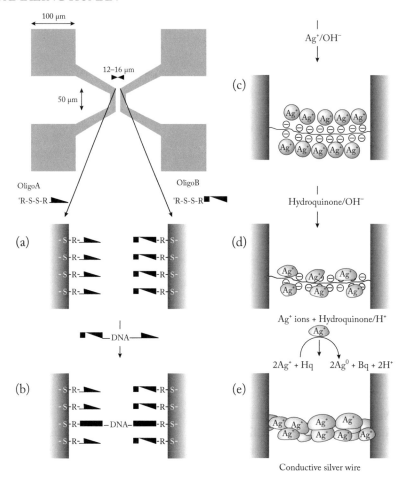

Figure 6.6: The process to create a silver wire using DNA as a template.

6.2 PROTEINS

The process of creating RNA from DNA is called transcription. After the RNA chain is complete the RNA, through the process of translation, results in proteins. There are multiple steps and nuances of the simplified process described in the last sentence, and the reality of going from the simplicity of four bases in DNA to the complexity of hundreds of thousands of possible proteins is nontrivial.

RNA consists of four bases: A, C, G, and U (uracil). Uracil has replaced thymine. RNA is a single-stranded structure. Groups of three bases, called codons, on the RNA strand will code for specific amino acids. There are 20 different amino acids. These amino acids are combined into strings to create a protein. The sequence of the amino acids determines the 3D structure and the specific function of the protein. Amino acids are charged entities with regions of positive and negative charge, and each amino acid will be different. The result is that as all the amino acids form into a polypeptide chain: ionic bonds, hydrogen bonds and forces such as van der Waals will have an impact on the polypeptide chain structure.

It is important to be aware of the vast number of possible proteins. Consider a protein, a polypeptide chain, that is to be composed of four amino acids. Assuming that an amino acid could be used more than once, there are 160,000 possible proteins that could be made from those four amino acids. Proteins are comprised of many more amino acids than just four, and therefore result in very complex structures with many turns, twists, and folds.

The protein structure is categorized into four levels of structure, as presented in Figure 6.7. The primary structure is merely the long chain of amino acids. The next level, or secondary structure, is dependent on the interactions between the amino acid component, in particular the interactions of the N-H groups and the C = O groups with each other and the orientation of the associated strands. This results in either an α-helix or a β-sheet structure. The third level, the tertiary structure, represents the overall shape. This 3D structure is determined by polar, nonpolar acid and R group charges that may exist. For proteins with multiple polypeptide chains, the subunits will join to form a quaternary level structure.

Included in the hundreds of thousands of possible protein structures lies equal number of functions. Table 6.1 lists the different types of proteins, examples of those types and the functions. Clearly, proteins play a major role in every aspect of a living organism.

The complexity inherent in the structure and composition of proteins allows for the multitude of functional capabilities. But it also can allow for unintended consequences. In many cases, there are fail safe, redundant or back-up approaches that compensate for errors that may occur in a given protein, i.e., a misplaced amino acid or two.

Understanding and defining the protein structure and how other substances will interact with that protein is critical in the discovery and development of pharmaceuticals. Even if the protein is pristine in its composition and structure, the drug molecule may be designed in such a way that it interacts with an unintended portion of the protein. This interaction is due to the charge distribution on the protein and the charge distribution on the drug molecule. An interaction that is not intended can result in one of the side effects often associated with a specific medication.

Therefore, it is important to be able to understand not only the physical structure of the protein but also the structure of the drug molecules. This is not an easy task: the amino acid length can range in size from 0.4–1.0 nm and many proteins are 3–6 nm in size. The collagen protein molecule is only 1.4 nm in diameter but 280 nm in length due to its triple helical structure. When considering the function of collagen, the length is appropriate. An antibody on the other hand is about 10 nm in length. Drug molecules are in the same size range of 3–10 nm. An aspirin molecule is on the small end of that scale with a molecular radius of 1 nm.

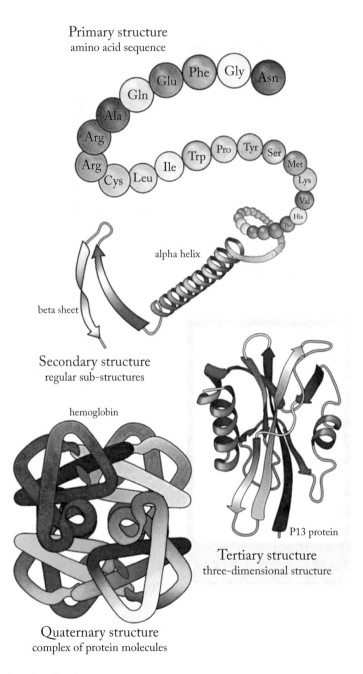

Figure 6.7: The four levels of a protein structure.

Table 6.1: Different types of proteins and their function

Protein Types	Examples	Functions
Structure	Actin, myosin, collagen, elastin, keratin	Give tissues (bone, tendons, ligaments, cartilage, skin, muscles) strength and structure
Enzymes	Amylase, lipase, pepsin, lactase	Digest macronutrients into smaller monomers that can be absorbed; performs steps in metabolic pathways to allow for nutrient utilization
Hormones	Insulin, glucagon, thyroxine	Chemical messengers that travel in blood and coordinate processes around the body
Fluid and acid–base balance	Albumin, hemoglobin	Maintains appropriate balance of fluids and pH in different body compartments
Transport	Hemoglobin, albumin, protein channels, carrier proteins	Carry substances around the body in the blood or lymph; help molecules cross cell membranes
Defense	Collagen, lysozyme, antibodies	Protect the body from foreign pathogens

One solution to the problem of capturing and characterizing individual molecules has been proposed by researchers in France [8]. The design includes tweezers made of strands of DNA. With one end of the DNA attached to a glass slide and the other end able to move up and down using a magnetic bead. The tweezer structure is shown in Figure 6.8.

It is critical to study the molecular binding dynamics, that is how the molecule will interact with other molecules. Molecules of the same chemical composition, may have different interactions because of their unique structure which results with different physical locations of charged regions, resulting in potentially unintended consequences of bonding to another molecule. This nuance is important in determining drug efficacy and side effects.

Not only are DNA and proteins astounding entities, in their own right, each with unique and multi-dimensional functions, but can work very well together to support further understanding.

6.3   REFERENCES

[1] Abu-Salah, K. M., Ansari, A. A., and Alrokayan, S. A. DNA-based applications in nanobiotechnology. *BioMed Research International*, 2010(715295), June 28, 2010. https://doi.org/10.1155/2010/715295 DOI: 10.1155/2010/715295.

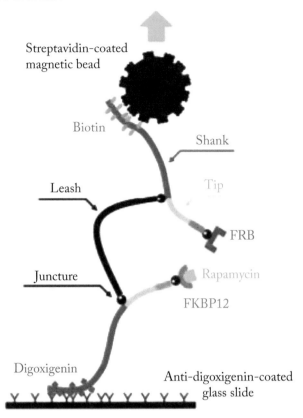

Figure 6.8: DNA tweezers for molecular study.

[2] Biologydictionary.net (Eds.) Nucleotide. *Biology Dictionary*, October 30, 2016. https://biologydictionary.net/nucleotide/

[3] Braun, E., Eichen, Y., Sivan, U. et al. DNA-templated assembly and electrode attachment of a conducting silver wire. *Nature*, 391:775–778, 1998. https://doi.org/10.1038/35826 DOI: 10.1038/35826.

[4] Callahan, A., Leonard, H., and Powell, T. *Nutrition: Science and Everyday Application*. PressBooks, 2020. https://openoregon.pressbooks.pub/nutritionscience/chapter/6b-protein-functions/

[5] Chen, T., Ren, L., Liu, X., Zhou, M., Li, L., Xu, J., and Zhu, X. DNA nanotechnology for cancer diagnosis and therapy. *International Journal of Molecular Sciences*, 19(6):1671, 2018. https://doi.org/10.3390/ijms19061671

[6] Gentile, F., Moretti, M., Limongi, T., et al. Direct imaging of DNA fibers: The visage of double helix. *Nano Letters*, 12(12):6453–6458, 2012. https://doi.org/10.1021/nl3039162

[7] DNA and RNA. https://www.coursehero.com/sg/cell-biology/dna-and-rna/

[8] Kostrz, D., Wayment-Steele, H. K., Wang, J. L., et al. A modular DNA scaffold to study protein—protein interactions at single-molecule resolution. *Nature Nanotechnology*, 14:988–993, 2019. https://doi.org/10.1038/s41565--019-0542-7 133

[9] Molnar, C. and Gair, J. *Concepts of Biology*, 1st Canadian ed., Bccampus, 2015. https://opentextbc.ca/biology/

[10] Panawala, L. Difference between purines and pyrimidines, 2017. https://www.researchgate.net/publication/316698909_Home_Science_Biology_Molecular_Biology_Difference_Between_Purines_and_Pyrimidines_Difference_Between_Purines_and_Pyrimidines

[11] Smith, Y. Protein structure and function. *News-Medical*, October 15, 2021. https://www.news-medical.net/life-sciences/Protein-Structure-and-Function.aspx.

Author's Biography

DEB NEWBERRY

Deb Newberry has been involved with research, emerging technology, and nanotechnology for several decades. She served as the Director/Instructor of the Nanoscience Technician program at Dakota County Technical College in Rosemount MN from 2004–2018. She created the 72-credit nanoscience technician program in 2003 and began the program with National Science Foundation funding. Deb also served as the Director and Principal Investigator of the Center for Nanotechnology Education, Nano-Link, for over 10 years. Nano-Link has been funded by over $12M from the National Science Foundation. Nano-Link educational content, developed by Deb, has been used by over 900 educators and has reached over 100,000 students. She is a nanotechnology book author and co-author, has written over 12 book chapters, is an IEEE Distinguished Lecturer, and presented more than 250 presentations and tens of educator workshops.

Deb Newberry is currently serving as CEO of Newberry Technology Associates (NTA), a company that provides expertise in organizational structure and efficiency, emerging technologies, strategic planning, technology evolution, and product development. The NTA team performs business and competitive analysis for multiple technologies including nanotechnology, electronics, photonics, material science, additive manufacturing, and biotechnology. The strategic focus is to determine how emerging technologies may impact products, companies, and market segments. Prior to her career in education, Deb worked in the corporate world for 24 years performing thermal and radiation testing and analysis on satellite systems and then serving as Executive Director managing over $450M in satellite programs.

Ms. Newberry led a national committee with a focus on determining the impact of nanotechnology on satellite electronics and has served on multiple advisory boards as well as state and national commissions. She is a member of professional organizations such as the IEEE, ACS, MRS, RCS, and ASEE and serves on multiple conference planning committees.

Printed in the United States
by Baker & Taylor Publisher Services